普通高等教育"计算机类专业"规划教材

Java程序设计实用教程习题集

高飞 赵小敏 陆佳炜 徐俊 编著

清华大学出版社
北京

内 容 简 介

本书主要包括了Java语言概述,Java基础语法,类和对象,类的封装性、继承性、多态性及接口,数组、字符串和枚举,Java常用类及接口,异常处理,流和文件,图形用户界面编程,多线程,网络编程,数据库编程,XML及程序打包等对应章节的习题,同时罗列了各章重要的知识点,在给出习题前还包括了习题的例题讲解,整个习题类型包括判断题、选择题、程序阅读题(含程序填空题)和编程题,其中部分选择题是多选。本书是《Java程序设计实用教程》的配套习题集。

本书可以作为高等院校计算机科学与技术专业、软件工程专业及相关专业的实验教材,也可作为自学教材。

本书封面贴有清华大学出版社防伪标签,无标签者不得销售。
版权所有,侵权必究。举报: 010-62782989, beiqinquan@tup.tsinghua.edu.cn。

图书在版编目(CIP)数据

Java程序设计实用教程习题集 / 高飞等编著. —北京: 清华大学出版社,2013(2024.10重印)
普通高等教育"计算机类专业"规划教材
ISBN 978-7-302-32051-7

Ⅰ. ①J… Ⅱ. ①高… Ⅲ. ①JAVA语言—程序设计—高等学校—习题集 Ⅳ. ①TP312-44

中国版本图书馆CIP数据核字(2013)第078647号

责任编辑:白立军
封面设计:常雪影
责任校对:时翠兰
责任印制:曹婉颖

出版发行:清华大学出版社
网　　址: https://www.tup.com.cn, https://www.wqxuetang.com
地　　址: 北京清华大学学研大厦A座　　邮　编: 100084
社 总 机: 010-83470000　　　　　　　　　邮　购: 010-62786544
投稿与读者服务: 010-62776969, c-service@tup.tsinghua.edu.cn
质 量 反 馈: 010-62772015, zhiliang@tup.tsinghua.edu.cn
课 件 下 载: https://www.tup.com.cn,010-83470236
印 装 者: 涿州市般润文化传播有限公司
经　　销: 全国新华书店
开　　本: 185mm×260mm　　印 张: 14　　字 数: 322千字
版　　次: 2013年6月第1版　　　　　　　印 次: 2024年10月第14次印刷
定　　价: 49.00元

产品编号: 052603-03

《Java 程序设计实用教程习题集》 前 言

  本书是《Java 程序设计实用教程》的配套习题集。作者团队从 2003 年开始在浙江工业大学软件学院开设了 Java 程序设计课程,至今每年都为浙江工业大学计算机科学与技术学院、软件学院的本科生讲授 Java 程序设计及相关课程。在教学过程中,作者团队自行设计了大量习题,用来课堂讲解、课后作业和期末考试。习题的收集与编写经历了很长一段时间,习题集主要由高飞、赵小敏、陆佳炜、徐俊等设计与编写,其中部分是从参考文献中的教材和习题集以及网络中收集并整理而来。在编写的过程中,史雪松等参与了部分编程题的代码设计、编写和测试工作,也有一些编程题参考代码来自于夏梦婕、吴晓蓓等同学上交的作业。编程题序号右上角打 * 的题目来自于为本科生设计的自主实验,参考代码来自于张春梅、李敏、陆阳、王雨晨和杜燕华等同学提交的源代码和实验报告。本书中的所有程序源代码和编程题参考代码,都经作者团队在 JDK 6 环境下由 Eclipse 开发工具编译通过和运行测试。书中所有练习题都给出了参考答案,部分容易出错的题目给出了解析。因篇幅有限,一些编程题参考代码不附在书上。书中的程序源代码和编程题参考代码可从本书配套网站(http://www.zjut-java.com)或清华大学出版社网站(http://www.tup.com.cn)下载。

  本书在习题收集、编写与出版的过程中得到了许多老师、同学和朋友的帮助,尤其得到了家人的支持,这里一并对他们表示诚挚的感谢。由于水平所限,时间紧迫,书中难免存在一些错误和缺陷,恳切希望广大读者特别是讲授 Java 程序设计课程的老师和学习 Java 的同学批评指正。也恳切希望能够通过 E-mail(feig@zjut.edu.cn)得到关于习题集的建议或意见。

高 飞

2013 年 1 月

第一部分　习题解析与练习题

第1章　Java语言概述　/3
- 1.1　知识点　/3
- 1.2　例题　/3
- 1.3　练习题　/4
  - 1.3.1　判断题　/4
  - 1.3.2　选择题　/5
  - 1.3.3　简答题　/7

第2章　Java基础语法　/8
- 2.1　知识点　/8
- 2.2　例题　/8
- 2.3　练习题　/11
  - 2.3.1　判断题　/11
  - 2.3.2　选择题　/11
  - 2.3.3　程序阅读题　/17
  - 2.3.4　编程题　/18

第3章　类和对象　/20
- 3.1　知识点　/20
- 3.2　例题　/20
- 3.3　练习题　/26
  - 3.3.1　判断题　/26
  - 3.3.2　选择题　/26
  - 3.3.3　程序阅读题　/32
  - 3.3.4　编程题　/36

第4章　类的封装性、继承性、多态性及接口　/39
- 4.1　知识点　/39
- 4.2　例题　/39
- 4.3　练习题　/50

    4.3.1 判断题 /50
    4.3.2 选择题 /51
    4.3.3 程序阅读题 /59
    4.3.4 编程题 /66

第5章 数组、字符串和枚举 /70
  5.1 知识点 /70
  5.2 例题 /70
  5.3 练习题 /72
    5.3.1 判断题 /72
    5.3.2 选择题 /73
    5.3.3 程序阅读题 /79
    5.3.4 编程题 /81

第6章 Java常用类及接口 /83
  6.1 知识点 /83
  6.2 例题 /83
  6.3 练习题 /86
    6.3.1 判断题 /86
    6.3.2 选择题 /86
    6.3.3 程序阅读题 /87
    6.3.4 编程题 /87

第7章 异常处理 /89
  7.1 知识点 /89
  7.2 例题 /89
  7.3 练习题 /91
    7.3.1 选择题 /91
    7.3.2 程序阅读题 /93
    7.3.3 编程题 /98

第8章 流和文件 /99
  8.1 知识点 /99

8.2 例题 /99
8.3 练习题 /106
　　8.3.1 判断题 /106
　　8.3.2 选择题 /106
　　8.3.3 程序阅读题 /108
　　8.3.4 编程题 /111

第 9 章 图形用户界面编程 /114
9.1 知识点 /114
9.2 例题 /114
9.3 练习题 /117
　　9.3.1 判断题 /117
　　9.3.2 选择题 /117
　　9.3.3 程序阅读题 /119
　　9.3.4 编程题 /125

第 10 章 多线程 /128
10.1 知识点 /128
10.2 例题 /128
10.3 练习题 /131
　　10.3.1 判断题 /131
　　10.3.2 选择题 /131
　　10.3.3 程序阅读题 /134
　　10.3.4 编程题 /139

第 11 章 网络编程 /141
11.1 知识点 /141
11.2 例题 /141
11.3 练习题 /144
　　11.3.1 判断题 /144
　　11.3.2 选择题 /145
　　11.3.3 程序阅读题 /146
　　11.3.4 编程题 /148

第12章　数据库编程　/150
- 12.1　知识点　/150
- 12.2　例题　/150
- 12.3　练习题　/154
  - 12.3.1　选择题　/154
  - 12.3.2　编程题　/156

第13章　XML及程序打包　/161
- 13.1　知识点　/161
- 13.2　例题　/161
- 13.3　练习题　/161
  - 13.3.1　判断题　/161
  - 13.3.2　选择题　/162

第二部分　参考答案

第14章　Java语言概述参考答案　/165
- 14.1　判断题　/165
- 14.2　选择题　/165
- 14.3　简答题　/165

第15章　Java基础语法参考答案　/167
- 15.1　判断题　/167
- 15.2　选择题　/167
- 15.3　程序阅读题　/167
- 15.4　编程题　/167

第16章　类和对象参考答案　/174
- 16.1　判断题　/174
- 16.2　选择题　/174
- 16.3　程序阅读题　/174

16.4 编程题 /175

第 17 章 类的封装性、继承性、多态性及接口参考答案 /180

17.1 判断题 /180
17.2 选择题 /180
17.3 程序阅读题 /181
17.4 编程题 /182

第 18 章 数组、字符串和枚举参考答案 /190

18.1 判断题 /190
18.2 选择题 /190
18.3 程序阅读题 /191
18.4 编程题 /191

第 19 章 Java 常用类及接口参考答案 /196

19.1 判断题 /196
19.2 选择题 /196
19.3 程序阅读题 /196
19.4 编程题 /196

第 20 章 异常处理参考答案 /198

20.1 选择题 /198
20.2 程序阅读题 /198
20.3 编程题 /198

第 21 章 流和文件参考答案 /202

21.1 判断题 /202
21.2 选择题 /202
21.3 程序阅读题 /202
21.4 编程题 /203

第 22 章　图形用户界面编程参考答案　/206
22.1　判断题　/206
22.2　选择题　/206
22.3　程序阅读题　/206
22.4　编程题　/207

第 23 章　多线程参考答案　/208
23.1　判断题　/208
23.2　选择题　/208
23.3　程序阅读题　/208
23.4　编程题　/210

第 24 章　网络编程参考答案　/211
24.1　判断题　/211
24.2　选择题　/211
24.3　程序阅读题　/211
24.4　编程题　/211

第 25 章　数据库编程参考答案　/212
25.1　选择题　/212
25.2　编程题　/212

第 26 章　XML 及程序打包参考答案　/213
26.1　判断题　/213
26.2　选择题　/213

参考文献　/214

第一部分
习题解析与练习题

这部分包括各章节的知识点以及重点和难点,例题与解析和练习题。知识点是要求学生应该要掌握的内容,重点是教学过程中着重讲解的内容,也是学生必须掌握的内容。例题是作者团队在教学过程中课堂上给学生讲解的例子,解析是分析题意,并给出理解或编写Java语言规范和程序应注意的地方。练习题涉及的题型包括判断题、选择题、程序阅读题(含程序填空题)以及编程题,这里的选择题部分是多选的。

第一部分

引起腐蚀的条件

...

第 1 章　Java 语言概述

1.1　知　识　点

（1）Java 的发展史。
（2）Java 的主要特征。
（3）Java 开发环境的配置。
（4）Java 程序的开发过程和工作原理。
（5）Java 的常用开发工具。
重点：Java 的特点、Java 程序的开发过程和工作原理、开发工具。
难点：Java 程序的开发过程和工作原理。

1.2　例　　题

【例 1-1】　Java 程序的执行过程中用到一套 JDK 工具，其中 java.exe 是指（　　）。
A. Java 文档生成器　　　　　　　　B. Java 解释器
C. Java 编译器　　　　　　　　　　D. Java 类分解器
【解析】　选 B。Java 文档生成器为 javadoc.exe，Java 解释器为 java.exe，Java 编译器为 javac.exe。

【例 1-2】　Java 语言具有许多优点和特点，下列选项中（　　）反映了 Java 程序并行机制的特点。
A. 安全性　　　　B. 多线程　　　　C. 跨平台　　　　D. 可移植
【解析】　选 B。Java 语言有下面一些特点：简单、面向对象、分布式、解释执行、健壮、安全、体系结构中立、可移植、高性能、多线程以及动态性。其中，多线程反映了 Java 程序并行机制的特点。

【例 1-3】　Java 程序分为哪两大类，它们之间有哪些相同点？主要区别有哪些？
【解析】　Java 程序分为 Java Application 和 Java Applet。
相同点如下所示。
（1）两类程序都由一个或多个以 class 为后缀的文件组成。
（2）两类程序都需要用户系统安装 Java 虚拟机（JVM）。Java 虚拟机能够载入并翻译 Java 程序，并且可以提供 Java 内核包的实现。
区别如下所示。
（1）Java Applet 程序可以被嵌入 HTML 网页内，从而可以在网络上发布，当网页被浏览时它们可以在浏览器中运行；而 Java Application 程序却不支持网页嵌入和下载。
（2）Java Applet 程序只能在与 Java 兼容的容器中运行，例如，网络浏览器；而 Java Application 程序却没有这个限制。

（3）一个 Java Applet 程序必须定义一个 Applet 类的子类，一个 Java Application 程序也可以定义一个 Applet 类的子类，但这不是必需的，一个 Java Application 程序必须在一个类中定义一个 main 方法，而一个 Java Applet 程序并不定义 main 方法，它的执行是由 Applet 类定义的多个方法控制的。

【例 1-4】 开发与运行 Java 程序需要经过哪些主要过程？

【解析】 开发与运行 Java 程序主要包括源程序编辑、字节码编译和解释运行等过程。首先，用文本编辑器或 JCreator 或 Eclipse 等 Java 开发工具编辑源程序，并保存；其次，用 Java 编译器工具 javac.exe 编译源程序文件，生成字节码.class 文件；最后，用 Java 解释器工具 java.exe 解释运行生成的.class 文件。

【例 1-5】 简述 Java 程序运行的过程。

【解析】 （1）类加载器(Class Loader)加载程序运行所需的所有类，它通过区分本机文件系统的类和网络系统导入的类而增加安全性，这可以限制任何木马程序，因为本机类总是先被加载，一旦所有的类被加载完，执行文件的内容划分就固定了，在这个时候特定的内存地址被分配给对应的符号引用，查找表(Lookup Table)也被建立，由于内存划分发生在运行时，解释器在受限制的代码区增加保护防止未授权的访问。

（2）字节码校验器(Byte Code Verifier)进行校验，主要执行下面的检查：类符合 JVM 规范的类文件格式，没有违反访问限制，代码没有造成堆栈的上溢或者下溢，所有操作代码的参数类型都是正确的，没有非法的数据类型转换发生。

（3）校验通过的字节码被解释器(Interpreter)执行，解释器在必要时通过运行时系统执行对底层硬件的合适调用。

1.3 练 习 题

1.3.1 判断题

1. Java 语言是一种解释执行的语言，这个特点是 Java 语言的一个缺点。 （ ）
2. Java 源程序的文件名一定要与文件中某个类的名称一致。 （ ）
3. Java 语言采用面向对象的思想编程，具有跨平台、分布式、多线程等优点。 （ ）
4. Java 中的标识符和关键字都是区分大小写的，如果把声明类的 class 写成 Class 或者 CLASS，编译会出错。 （ ）
5. Java 源程序编写好之后，以文件的形式保存在硬盘或 U 盘上，源文件的名字可以随便取，它不一定与程序的主类名一致。 （ ）
6. 在 JDK 命令行开发工具中，用编译程序 javac.exe 编译生成的文件是二进制可执行文件。 （ ）
7. Java 程序可以分为 Java Application 和 Java Applet 两类。 （ ）
8. Java 虚拟机可以在 Microsoft Windows 系列、Linux、UNIX、Mac OS 等操作系统下执行 Java 字节码程序。 （ ）
9. javac 是 Java 编译器，用来将 Java 源程序编译成 Java 字节码的程序。字节码文件名和源文件名相同，扩展名是 class。 （ ）

10. appletviewer 是 Java 小应用程序(Applet)浏览器,可以执行 HTML 文件中的 Java Applet。它是个模拟浏览器,可以显示 Applet 的运行结果。   (   )

1.3.2 选择题

1. Java 工具 JDK 中用来运行 Applet 程序的命令是(   )。
   A. java   B. javac   C. applet   D. appletviewer
2. 下列说法不正确的是(   )。
   A. 一个 Java 源程序经过编译后,得到的文件扩展名一定是 class
   B. 一个 Java 源程序编译通过后,得到的结果文件数也只有一个
   C. 一个 Java 源程序编译通过后,得到的结果文件数可能有多个
   D. 一个 Java 源程序编译通过后,不一定能用 Java 解释器执行
3. 编译 Java 源程序文件将产生相应的字节码文件,这些字节码文件的扩展名为(   )。
   A. java   B. class   C. html   D. exe
4. 安装 JDK 时,为了能方便地编译和运行程序,应该设置环境变量,其中主要的环境变量的名称是(   )。
   A. JAVAHOME   B. java   C. path   D. classpath
5. Java 编译程序的文件名是(   )。
   A. java.exe   B. javadoc.exe   C. javac.exe   D. jar.exe
6. 下列说法中正确的是(   )。
   A. Java 是不区分大小写的,源文件名与程序类名不允许相同
   B. Java 语言以函数为程序的基本单位
   C. Applet 是 Java 的一类特殊应用程序,它可嵌入 HTML 中发布到互联网上
   D. 以//符号开始的为多行注释语句
7. 下面说法中正确的是(   )。
   A. Java 程序的源文件名称是与主类的名称相同,后缀可为.java 或.txt 等
   B. JDK 的编译命令是 java
   C. 一个 Java 源程序编译后可能产生几个字节码文件
   D. 在 DOS 命令行下编译好字节码文件后,只需直接输入程序名即可运行该程序
8. 下面有关 Java 代码安全性的叙述错误的是(   )。
   A. 字节码校验器加载查询执行所需的所有类
   B. 运行时,由解释器执行代码
   C. 运行时,字节码被加载、验证,然后在解释器里运行
   D. 类加载器通过分离本机文件系统的类和从网络导入的类增加安全性
9. Java 的主要优点是(   )。
   A. 直接操作内存,功能强大
   B. 一次编译,到处运行
   C. 纯面向对象的语言

D. 可以通过拖曳的方式快速开发程序界面

10. Java 应用程序执行入口的 main() 方法返回类型是（　　）。
    A. int　　　　　B. void　　　　　C. boolean　　　　　D. static

11. 某 Java 程序的类定义如下：

    public class MyClass extends BaseClass{ }

    则该 Java 源文件在存盘时的源文件名应为（　　）。
    A. myclass.java　　　　　　　　B. MyClass.java
    C. MYCLASS.java　　　　　　　D. MyClass.class

12. 在一个合法的 Java 源程序文件中定义了 3 个类，其中属性为 public 的类可能有（　　）个。
    A. 0　　　　　B. 1　　　　　C. 2　　　　　D. 3

13. 编译一个定义了两个类和三个方法的 Java 源程序文件，总共会产生（　　）个字节码文件，这些字节码文件的扩展名是（　　）。
    A. 2，以 class 为扩展名　　　　B. 2，以 java 为扩展名
    C. 5，以 class 为扩展名　　　　D. 5，以 java 为扩展名

14. 设 HelloWorld.java 的代码如下：

    ```
    class HelloWorld{
        public void main(String a[]){
            System.out.println("Hello World!");
        }
    }
    ```

    下面说法正确的是（　　）。
    A. HelloWorld.java 无法通过编译，因为 main 方法的声明方式不对
    B. HelloWorld.java 可以通过编译，但运行该程序会出现异常，不会打印 Hello World!
    C. HelloWorld.java 可以通过编译，但无法运行，因为该文件没有 public 类
    D. HelloWorld.java 可以通过编译并正常运行，结果输出 Hello World!

15. 若 Java 的安装目录是 C:\Java\jdk1.6，则为了能够方便地使用 javac.exe 编译 Java 程序，应该进行下列哪一项环境变量的设置？（　　）
    A. 编辑环境变量 path，在其变量值的尾部增加";C:\Java\jdk1.6\bin;"
    B. 编辑环境变量 path，在其变量值的尾部增加";C:\Java\jdk1.6\jre;"
    C. 编辑环境变量 path，在其变量值的尾部增加";C:\Java\jdk1.6\lib;"
    D. 编辑环境变量 path，在其变量值的尾部增加";C:\Java\jdk1.6\include;"

16. 在 DOS 命令行状态下，如果源程序 HelloWorld.java 在当前目录下，那么编译该程序的命令是（　　）。
    A. java HelloWorld　　　　　　B. java HelloWorld.java
    C. javac HelloWorld　　　　　　D. javac HelloWorld.java

17. 在 DOS 命令行状态下,如果命令 java Hello 成功运行了程序,那么下面哪些叙述是正确的?（　　）

    A. 当前目录中一定存在文件 Hello.java

    B. 类 Hello 中一定含有 main 方法

    C. 当前目录中一定存在文件 Hello.class

    D. 当前目录中可以不存在文件 Hello.java

1.3.3 简答题

1. 简述 Java 程序的可移植性。
2. Java 程序是由什么组成的？Java 源文件的命名规则是怎样的？

第 2 章　Java 基础语法

2.1　知　识　点

（1）标识符和关键字。
（2）基本数据类型和类型转换。
（3）常量和变量。
（4）运算符：算术运算符、关系运算符、逻辑运算符、位运算符、赋值类运算符、条件运算符、对象运算符等。
（5）程序流程控制语句(if-else、switch、while、do-while、for、break、continue、return)和注释语句。
（6）输入参数方式。

重点：基本数据类型和类型转换，常量和变量，运算符，程序流程控制语句。

难点：数据类型转换，程序流程控制语句。

2.2　例　　题

【例 2-1】　在 Java 程序中，下面哪个是不合法的标识符？（　　）
A. 2D　　　　　　B. True　　　　　　C. _name　　　　　　D. T1

【解析】　选 A。Java 中的标识符可以用来标识变量名、类名、类中的方法名和文件名等。合法的标识符由字母、数字、下划线"_"和美元符号 $ 组成，其中首字符必须是字母、下划线"_"或美元符号 $。另外，Java 关键字 false、true 和 null 等也不能作为标识符。

【例 2-2】　下面哪个定义变量的语句不合规范？（　　）
A. int retireAge=60;
B. final int RETIRE_AGE=60;
C. static int retireAge=60;
D. private int RETIREAGE=60;

【解析】　选 D。Java 标识符的习惯命名规范如下。
（1）用能表达明确意义的英文单词，并采用规范的单词缩写形式与单词分隔形式。
（2）表示常量时标识符全部用大写字母和下划线表示，如 PI、SALES_TAX。
（3）表示类名或接口名时，标识符用大写字母开头，如 CreditCard。
（4）表示变量名和方法名，以小写字母开头，单词之间不要有分隔符，第二及后面单词第一个字符用大写字母，如 authorName。

【例 2-3】　下面哪个赋值语句会产生编译错误？（　　）
A. float a=1.0;
B. double b=1.0;
C. int c=1+'1';
D. long d=1;

【解析】 选 A,应改为"float a=1.0F;"或"float a=1.0f;"。

Java 的基本数据类型包括 boolean、char、byte、short、int、long、float、double,每种数据类型有默认值和封装类,如表 2-1 所示。

表 2-1 各种数据类型的默认值和封装类

类型	默认值	封装类
boolean	false	Boolean
char	'\u0000'(空字符)	Char
byte	(byte)0	Byte
short	(short)0	Short
int	0	Integer
long	0L 或 0l	Long
float	0.0f 或 0.0F	Float
double	0.0 或 0.0d 或 0.0D	Double

【例 2-4】 用 for 语句编写程序实现求 1+2+…+100 的和的实例。

参考代码如下:

```
1.  public class ForOfSum{
2.      public static void main(String[ ] args)  {
3.          int sum=0;                         //累加器 sum 初始值为 0
4.          for(int i=1;i<=100;i++){
5.              sum+=i;
6.          }
7.          System.out.println("sum="+sum);
8.      }
9.  }
```

【解析】 编写 for 语句时,需避免以下两种常见的错误。

错误一:for 语句不写大括号{ },导致以下第 3 行代码无法找到变量 i 和 sum。

```
1. for(int i=1,sum=0;i<=100;i++)
2.     sum+=i;
3.     System.out.println("i="+i+",+sum="+sum);
```

错误二:利用浮点类型的==或!=运算作为条件表达式。

```
1. for(double x=0.01;x!=1.0;x+=0.01){
2.     System.out.println(x+";");
3. }
```

由于浮点类型的算术运算都是近似计算,可能导致以上 for 语句无限循环执行。

【例 2-5】 用 switch 语句实现学生成绩的百分制到五级等级制转换的实例。

参考代码如下:

```
1. class SwitchDemo{
2.     public static void main(String[] args) {
3.         int testScore=88;
4.         char grade;
5.         switch (testScore/10) {             //两个整型数相除的结果还是整型
6.             case 10:                         //此处没有使用 break
7.             case 9: grade='A'; break;        //值为 10 和 9 时的操作是相同的
8.             case 8: grade='B'; break;
9.             case 7: grade='C'; break;
10.            case 6: grade='D'; break;
11.            default:grade='F'; break;
12.        }
13.        System.out.println("grade is:"+grade);
14.    }
15. }
```

【解析】 使用 switch 语句时,应注意以下几点:

(1) 使用 switch 语句时,要注意表达式必须是符合 byte、char、short、int 或枚举类型的表达式,而不能使用浮点类型或 long 类型,也不能为一个字符串。

(2) switch 语句将表达式的值依次与每个 case 子句中的常量值相比较。如果匹配成功,则执行该 case 子句中常量值后的语句,直到遇到 break 语句为止。

(3) case 子句中常量的类型必须与表达式的类型兼容,而且每个 case 子句中常量的值必须是不同的。

(4) default 子句是可选的,当表达式的值与任一 case 子句中的值都不匹配时,就执行 default 后的语句。

(5) break 语句在执行完一个 case 分支后,使程序跳出 switch 语句,执行 switch 语句的后续语句。

(6) 在一些特殊的情况下,例如,多个不同的 case 值要执行一组相同的操作,可以写成如下形式:

```
⋮
case 常量 n:
case 常量 n+1:语句
    [break;]
⋮
```

(7) case 分支中包括多个执行语句时,可以不用花括号{ }括起。

(8) 通过 if-else 语句可以实现 switch 语句所有的功能。但通常使用 switch 语句更简练,且可读性强,程序的执行效率也高。

(9) if-else 语句可以基于一个范围内的值或一个条件来进行不同的操作,但 switch 语句中的每个 case 子句都必须对应一个单值。

【例 2-6】 用 break 语句编写一个程序的实例。

参考代码如下:

```
1. public class BreakDemo {
2.     public static void main(String[ ] args) {
3.         int index=0;
4.         while (index<=100) {
5.             index+=10;
6.             /*当 index 的值大于 100 时,循环将终止。但有一种特殊的情况,如果 index 的值等于
                40,循环也将立即终止*/
7.             if (index==40)
8.                 break;
9.             System.out.println("The index is "+index);
10.        }
11.    }
12. }
```

【解析】 用 break 语句可以终止 while、do-while、for 等循环语句的正常执行,在这些循环语句中,一旦遇到 break 语句,则退出循环。但要注意的是 break 语句不能用于除循环语句或 switch 语句外的任何其他语句中。continue 和 break 语句的区别是 continue 语句只结束本次循环,而 break 语句则是结束整个循环语句的执行。

以上代码中的 break 语句作用于 4~10 行,运行输出结果如下:

```
The index is 10
The index is 20
The index is 30
```

2.3 练 习 题

2.3.1 判断题

1. 在 Java 的基本数据类型中,char 型占用 16 位,即 2 个字节的内存空间。（   ）
2. 因为布尔类型 boolean 与其他基本数据类型不可以互相转换,所以布尔类型与其他基本数据类型不存在强弱关系。（   ）
3. Java 提供了 3 类流程控制语句,即顺序结构、选择结构和循环结构。（   ）
4. 在 Java 语言中,若没有括号的情况下,if 和 else 的匹配采用最近原则,即 else 总是与离它最近的 if 进行配对。（   ）
5. float x=26f; int y=26; int z=x/y; 以上语句能正常编译和运行。（   ）

2.3.2 选择题

1. 下列标识符不合法的是(   )。
   A. $ variable   B. _variable   C. variable5   D. break
2. 下列(   )不属于 Java 的基本数据类型。
   A. int   B. String   C. double   D. boolean

3. 下列答案正确的是（   ）。
   A. int n＝7；int b＝2＊n＋＋；结果：b＝15，n＝8
   B. int n＝7；int b＝2＊n＋＋；结果：b＝16，n＝8
   C. int n＝7；int b＝2＊n＋＋；结果：b＝14，n＝8
   D. int n＝7；int b＝2＊n＋＋；结果：b＝14，n＝7

4. 下列答案正确的是（   ）。
   A. int n＝7；int b＝2；结果：n/b 的值为 3.5
   B. int n＝7；int b＝2；结果：n/b 的值为 3.5L
   C. int n＝7；int b＝2；结果：n/b 的值为 3.5D
   D. int n＝7；int b＝2；结果：n/b 的值为 3

5. 下列（   ）不能作为 switch 表达式的数据类型。
   A. int   B. char   C. short   D. long

6. 范围大的数据类型要转换成范围小的数据类型，需要用到（   ）类型转换。
   A. 隐式   B. 强制   C. 不需要   D. 强弱

7. System.out.print("1"＋2)打印到屏幕的结果是（   ）。
   A. 3   B. 12   C. 1＋2   D. 4

8. 下面（   ）是不合法的变量名称。
   A. while-true   B. True   C. name   D. T1

9. 下列变量定义正确的是（   ）。
   A. boolean status＝1;
   B. float d＝45.6;
   C. char ch＝"a";
   D. int k＝1＋'1';

10. 假设 int x＝4,y＝100，下列语句的循环体共执行（   ）次。
```
while(y/x>3){
    if(y%x>3) {
        x=x+1;
    }
    else{
        y=y/x;
    }
}
```
    A. 1   B. 2   C. 3   D. 4

11. 程序 Test.java 的代码如下：
```
public class Test{
  public static void main(String[] args){
    System.out.println(args[2]);
  }
}
```
    以上程序编译后用 java Test 1 2 3 4 运行的输出结果是（   ）。
    A. 1   B. 2   C. 3   D. 4

12. 某个 main()方法中有以下代码：

    ```
    String s1,s2;
    int[] numbers;
    int num;
    num=15;
    boolean switcher=false;
    ```

    下列哪个说法是正确的？（   ）

    A．声明了 1 个基本类型变量和 2 个引用变量

    B．声明了 2 个基本类型变量和 2 个引用变量

    C．声明了 2 个基本类型变量和 3 个引用变量

    D．声明了 3 个基本类型变量和 3 个引用变量

13. 考查下列程序代码：

    ```
    final int BASE=10;
    int nubmer;
    ```

    下列哪个表达式可以求出 number 的最后一位数字？（   ）

    A．number-BASE   B．BASE/number

    C．BASE％number   D．number％BASE

14. 某个 main()方法中有以下的声明：

    ```
    final int MIN=0;
    final int MAX=10;
    int num=5;
    ```

    下列哪个语句可以用来表示"num 的值大于等于 MIN 并且小于等于 MAX"？（   ）

    A．!(num<MIN && num>MAX)

    B．num>=MIN && num<=MAX

    C．num>MIN||num<=MAX

    D．num>=MIN||num<=MAX

15. 考查下面的程序代码

    ```
    int num1=40;
    int num2=5;
    int ans=0;
    if (num1/5==num2) { ans=10;  }
    if (num2%5==0) { ans=20;  }
    if (num2==0) { ans=30;  } else { ans=40;  }
    System.out.println("answer is: " +ans);
    ```

    System.out.println 语句将输出什么结果？（   ）

    A．answer is：30   B．answer is：20

    C．answer is：10   D．answer is：40

16. 下面的代码段输出什么内容？（   ）

```
int num;
for (int row=0; row<3; row=row+1) {
  for (int col=0; col<3; col=col+1) {
    num=(row * 3)+(col+1);
    System.out.print(num);
    System.out.print("-");
  }
  System.out.println();
}
```

A. 1-2-3-4-5-6-7-8-9-
   3-6-9-
   6-9-12-

B. 0-3-6-

C. 1-4-7-
   2-5-8-
   3-6-9-

D. 1-2-3-
   4-5-6-
   7-8-9-

17. 假设程序已有以下变量定义：

```
int num=60;
```

下面的代码段将输出什么内容？（   ）

```
if (num%3==0) {
  System.out.println("number is divisible by 3");
  if (num%2==0) {
      System.out.println("number is divisible by 2");
  }
}
else {
    if (num%5==0) {
      System.out.println("number is divisible by 5");
    }
}
```

A. number is divisible by 3
B. number is divisible by 5
C. number is divisible by 3
   number is divisible by 2
   number is divisible by 5
D. number is divisible by 3
   number is divisible by 2

18. 哪一项可以填充代码中的空白？（   ）

```
public class TestOR{
  public static void main(String[] args){
    int a=111111;
    int b=222222;
        //以下代码将 a 和 b 值互换
```

```
        System.out.println("a="+a+"    b="+b);
    }
}
```

    A. a＝a^b;　　b＝a^b;　　a＝a^b;　　　　B. a＝a^b;　　a＝a^b;　　b＝a^b;
    C. b＝a^b;　　b＝a^b;　　a＝a^b;　　　　D. b＝a^b;　　a＝a^b;　　a＝a^b;

19. 以下代码的运行结果是(　　)。

```
import java.lang.Math;
public class Count {
    public static void main(String args[]) {
        int arr[]=new int [1000];
        int count[]=new int [21];
        int i,j;
        int all=0;
        for(i=0;i<1000;i++){
            arr[i]=(int)(Math.random()*20+0.5);
        }
        for(i=0;i<21;i++){
            for(j=0;j<1000;j++){
                if(arr[j]==i)
                    count[i]++;
            }
            all+=count[i];
        }
        System.out.println(all);
    }
}
```

    A. 100　　　　　　B. 1000　　　　　　C. 999　　　　　　D. 1001

20. 定义变量 boolean b＝true;则 String.valueOf(b)的类型是(　　)。
    A. boolean　　　　B. String　　　　　C. char　　　　　　D. int

21. 以下哪个不是 Java 关键字?(　　)
    A. void　　　　　　B. sizeof　　　　　C. const　　　　　　D. super

22. 下面哪个不是 Java 的基本数据类型?(　　)
    A. short　　　　　　B. Boolean　　　　C. byte　　　　　　D. float

23. int 的取值范围是(　　)。
    A. $-2^7 \sim 2^7-1$　　　　　　　　　　B. $0 \sim 2^{32}-1$
    C. $-2^{15} \sim 2^{15}-1$　　　　　　　　　D. $-2^{31} \sim 2^{31}-1$

24. 给出下面代码:

```
if (x>0) {
  System.out.println("first");
}
else if (x>-3) {
```

```
        System.out.println("second");
    }
    else {
        System.out.println("third");
    }
```

x 的取值在什么范围内时将打印字符串"second"?（　　）

A. x＞0　　　　　　　　　　　B. x＞－3

C. x＜＝－3　　　　　　　　　D. x＜＝0 && x＞－3

25. 以下表达式语句合法的是（　　）。

    A. byte a＝128；　　　　　　B. Boolean b＝null；

    C. int c＝1.0；　　　　　　　D. float d＝0.9239；

26. 下列代码哪行会出错？（　　）

```
1.  public void modify() {
2.      int i, j, k;
3.      i=100;
4.      while ( i>0 ) {
5.          j=i*2;
6.          System.out.println(" The value of j is " +j );
7.          k=k +1;
8.          i--;
9.      }
10. }
```

    A. 第4行　　　B. 第6行　　　C. 第7行　　　D. 第8行

27. 以下表达式语句不合法的是（　　）。

    A. double a＝1.0；

    B. Double a＝new Double(1.0)；

    C. byte a＝340；

    D. Byte a＝120；

28. 设 int x＝1,y＝2,z＝3；则语句 y＋＝z－－/＋＋x－x；执行后 y 的值是（　　）。

    A. 0　　　　　　B. 1　　　　　　C. 2　　　　　　D. 3

29. 下面哪个赋值语句是不合法的？（　　）

    A. float f＝20.3；　　　　　　B. double d＝2.3E12；

    C. double d＝2.1352；　　　　D. double d＝3.14D

30. 下列语句的输出应该是（　　）。

```
int x=4;
System.out.println("value is "+((x>4)?88.8:8));
```

    A. 输出结果为：value is 88.8　　　B. 输出结果为：value is 8

    C. 输出结果为：value is 8.0　　　 D. 语句错误

31. 下列关于注释语句的描述中,错误的是(   )。
    A. 以//开始的是单行注释语句
    B. 以/*开始,*/结束的是单行注释语句
    C. 以/*开始,*/结束的是多行注释语句
    D. 以/**开始,*/结束的是可以用于生成帮助文档的多行注释语句
32. 以下的变量定义语句中,合法的是(   )。
    A. float a＝3.4;
    B. char c＝1＋'1';
    C. double ＄a＊5＝2.0D;
    D. String name＃2＝"john";
33. 以下代码段执行后的输出结果为(   )。

    ```
    int x=7;
    double y=-5.0;
    System.out.println(x%y);
    ```

    A. 2.0   B. －2.0   C. 0.4   D. －0.4
34. DOS命令行下成功执行命令:java Test abc 21,则以下描述正确的是(   )。
    A. 该命令向main方法传递了3个参数,3个参数都是字符串
    B. 该命令向main方法传递了3个参数,2个是字符串,1个是整数
    C. 该命令向main方法传递了2个参数,2个参数都是字符串
    D. 该命令向main方法传递了2个参数,1个是字符串,1个是整数
35. 以下程序的输出结果为(   )。

    ```
    public class Test {
      public static void main(String args[]) {
        int i=0;
        for (i=0;i<4;i++) {
          if (i==3)
            break;
          System.out.print(i);
        }
        System.out.println(i);
      }
    }
    ```

    A. 0123   B. 0122   C. 123   D. 234

2.3.3 程序阅读题

1. 仔细阅读下面的程序代码,回答问题。

    ```
    class TestChoose{
      public static void main(String args[]){
          int n=2;
          while(n<=10){
              boolean flag=true;
              int k1=2;
    ```

```
            while(k1<=n/2+1){
                if(n%k1==0){
                    flag=false;
                    break;
                }
                k1++;
            }
            if(flag){
                System.out.println(n+" ");
            }
            n++;
        }
    }
}
```

问：(1) 程序的功能是什么？

(2) 程序的运行结果是什么？

2. 写出以下程序代码的运行结果。

```
public class Test{
    public static void main(String args[]){
        for(int i=1; i<3; i++)  {
          for(int j=3; j>0; j--)  {
            if(i==j)
                break;
            if(i%j==0)
                continue;
            System.out.println("i="+i +" j="+j);
          }
        }
    }
}
```

2.3.4 编程题

1. 打印某学生某一学期的 Java、数据库和英语等课程的成绩和等级(优、良、中、及格和不及格)，并给出平均成绩。

2. 编写一个程序，打印 100～200 之间的素数，要求每行按 10 个数(数与数之间有一个空格间隔)的形式对其输出。

3. 编写一个程序，给定一个 t 的值(可初始化定义)，按下式计算 y 值并输出，要求分别写出 if 语句和 switch 语句。

$$y = \begin{cases} t^2 - 1 & 0 \leqslant t < 1 \\ t^3 - 2 \cdot t - 2 & 1 \leqslant t < 3 \\ t^2 - t \cdot \sin t & 3 \leqslant t < 5 \\ t + 1 & 5 \leqslant t < 7 \\ t - 1 & \text{其他} \end{cases}$$

4. 用 for 语句实现下面的程序 ForDemo.java：从 10～100 能被 2 整数但不能被 3 整除的整数，要求每行按 10 个数（数与数之间有一个空格间隔）的形式对其输出。

5. 编写一个 Java 应用程序，先产生 1 个随机数 n（要求 $0<n<10$），然后随机生成 n 个 0～100 的随机数，输出这 n 个随机数的和。

6. 姐妹素数是指相邻两个奇数均为素数，请编写一个程序 SisterPrime.java 找出 100～1000 的所有姐妹素数。

7. 打印出所有的"水仙花数"。"水仙花数"是指一个三位数，其各位数字的立方和等于该数本身。例如，153 是一个"水仙花数"，因为 $153=1^3+5^3+3^3$。

8. 求 1000 之内的所有完全数。完全数就是一个数恰好等于它的因子之和。例如，6 的因子为 1, 2, 3，而 $6=1+2+3$，因此 6 就是完全数。

9. 编程求出 $e=1+1/1!+1/2!+1/3!+\cdots+1/n!+\cdots$ 的近似值，当 $1/n!$ 小于 0.0001 时停止计算。程序的源文件名为 CalculateE.java。类名为 CalculateE（提示：注意数据类型）。

10. 编写一个 Java 应用程序，从键盘输入 x，利用下列泰勒公式计算 $\cos(x)$ 的值，并输出（要求精确到 0.000 001）。泰勒公式为：
$$\cos(x) = 1-x^2/2!+x^4/4!-x^6/6!+x^8/8!-\cdots$$

第 3 章 类 和 对 象

3.1 知 识 点

(1) 面向对象的基本概念和特征。

对象是客观世界中的某个具体事物,对象的概念是面向对象技术的核心。面向对象技术中的对象就是现实世界中某个具体的物理实体在计算机逻辑中的映射和体现,它可以是有形的,也可以是无形的。例如,电视是一个具体存在的,拥有外形、尺寸、颜色等外部特性(或称属性)和开、关、设置等功能的实体。从程序设计的角度来看,事物的属性或特性可以用变量来表示,行为或功能则用方法来表示。面向对象的程序设计方法就是将客观事物抽象成为"类",并通过类的"继承"实现软件的可扩充性和可重用性。

类是同种对象的集合与抽象。在面向对象的程序设计中,定义类的概念来表述同种对象的公共属性和特点。类是一种抽象的数据类型,它是具有一定共性的对象的抽象,而属于类的某一对象称为类的一个实例,是类的一次实例化的结果。

用面向对象程序设计解决实际问题的基本思想是:首先将实际存在的物理实体抽象成概念世界的抽象数据类型,这个抽象数据类型里面包括了实体中与需要解决的问题相关的数据和属性;然后再用面向对象的工具,如 Java 语言,将这个抽象数据类型用计算机逻辑表达出来,即构造计算机能够理解和处理的类;最后将类实例化就得到了现实世界实体的面向对象的映射——对象,在程序中对对象进行操作,就可以模拟现实世界中实际问题并解决之。

(2) 类的定义、成员变量与成员方法、构造方法。

(3) 对象的生成与使用、变量的作用域、对象的内存分配机制、方法参数的传递。

(4) 关键字:this、static 和 final。

重点:类的定义,对象的生成与使用,构造方法的定义与使用,变量的作用域,对象的内存分配机制,方法参数的传递,关键字 this、static 和 final 的使用方法。

难点:类与对象的概念、对象的生成与使用、对象的内存分配机制、方法参数的传递、关键字 this、static 和 final 的使用方法。

3.2 例 题

【例 3-1】 定义一个学生类的实例。

```
1.  public class Student {
2.    String name;
3.    char sex;
4.    int stuID;
5.    public Student(){ }
```

```
6.     public Student(String stuName,char sex,int stuID){
7.         name=stuName;
8.         this.sex=sex;
9.         this.stuID=stuID;
10.    }
11.    public void setName(String stuName){
12.        name=stuName;
13.    }
14.    public void setSex(char sex){
15.        this.sex=sex;
16.    }
17.    public void setStuID(int stuID){
18.        this.stuID=stuID;
19.    }
20. }
```

【解析】 类由类首声明、成员变量和方法组成,其中类首声明的格式为[<修饰符>] class <类名> [extends <超类名>] [implements <接口名>]。第1行为类首声明,第2至第4行为成员变量定义,第5行和第6行是构造方法,第11行、第14行和第17行是类的成员方法的定义。

【例 3-2】 对象的内存分配机制的实例。

```
1.  public class Shirt{
2.    char size;
3.    float price;
4.    boolean longSleeved;
5.    public static void main (String args[]){
6.        Shirt myShirt=new Shirt();
7.        Shirt anotherShirt=new Shirt();
8.        myShirt.size='M';
9.        myShirt.price=22.99F;
10.       myShirt.longSleeved=false;
11.       anotherShirt.size='L';
12.       anotherShirt.price=29.99F;
13.       anotherShirt.longSleeved=true;
14.       anotherShirt=myShirt;
15.       anotherShirt.size='S';
16.       System.out.println("myShirt的尺寸大小为:"+myShirt.size);
17.    }
18. }
```

【解析】 Java程序创建对象(或实例化)就是为对象分配内存存储空间,并对对象进行初始化的过程。用new运算符和类的构造方法来完成创建对象,格式为:<类名> <对象名>=new <类名>(参数)。main方法是Java程序的执行入口,本例的程序执行第6和第7行时,JVM申请内存空间,为Shirt的对象引用anotherShirt和myShirt分配内存空间,并将 Shirt 的成员变量初始化,值为成员变量类型的默认值,如图 3-1 所示。

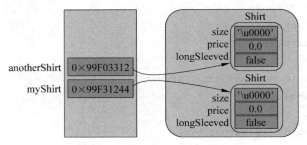

图 3-1　类 Shirt 的对象的内存空间(一)

程序执行第 8 行至第 10 行后,将 myShirt 对象内存空间中的 size 值修改为'M',price 值修改为 22.99F,longSleeved 值修改为 false,内存空间如图 3-2 所示。

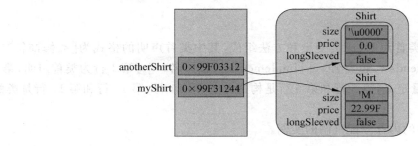

图 3-2　类 Shirt 的对象的内存空间(二)

程序执行第 11 行至第 13 行后,将 anotherShirt 对象内存空间中的 size 值修改为'L',price 值修改为 29.99F,longSleeved 值修改为 true,内存空间如图 3-3 所示。

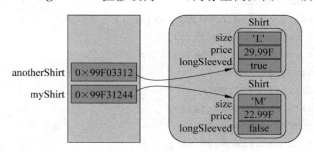

图 3-3　类 Shirt 的对象的内存空间(三)

程序执行第 14 行后,anotherShirt 对象指向 myShirt 对象的引用,anotherShirt 对象原有的内存空间因没有对象引用而成为垃圾内存空间,内存空间如图 3-4 所示。

程序执行第 15 行后,将 anotherShirt 对象内存空间中的 size 值修改为'S',内存空间如图 3-5 所示。

由于 myShirt 对象和 anotherShirt 对象引用同一个内存空间,使得 myShirt 的 size 值也为'S'。程序执行第 16 行后,输出结果为 myShirt 的尺寸大小为 S。

【例 3-3】　对象作为方法参数传递的实例。

```
1.  class ValObject{
```

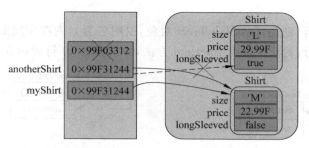

图 3-4 类 Shirt 的对象的内存空间(四)

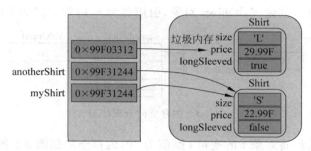

图 3-5 类 Shirt 的对象的内存空间(五)

```
2.     public int i=10;
3.   }
4.   public class TestValObject{
5.     public static void main(String argv[]){
6.       TestValObject t=new TestValObject();
7.       t.amethod();
8.     }
9.     public void amethod(){
10.      int i=20;
11.      ValObject v=new ValObject();
12.      v.i=30;
13.      another(v , i);
14.      System.out.print(v.i);
15.    }
16.    public void another(ValObject v, int i){
17.      i=0;
18.      v.i=40;
19.      ValObject vo=new ValObject();
20.      v=vo;
21.      System.out.print(v.i);
22.      System.out.print(i);
23.    }
24.  }
```

【解析】 程序从 main 方法入口开始执行,具体执行过程如下(为解释方便,将内存空间的地址省略)。

(1) 执行第 6 行,创建 TestValObject 对象,引用名为 t,内存空间如图 3-6 所示。
(2) 执行第 7 行,对象 t 调用 amethod 方法,并执行第 10 行后内存空间如图 3-7 所示。

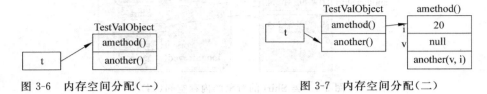

图 3-6 内存空间分配(一)　　　　　　图 3-7 内存空间分配(二)

(3) 执行第 11 行,创建 ValObject 对象,引用名为 v,内存空间如下图 3-8 所示。

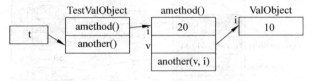

图 3-8 内存空间分配(三)

(4) 执行第 12 行,将对象 t 的变量 i 赋值为 30,内存空间如图 3-9 所示。

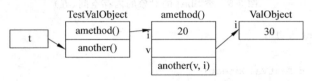

图 3-9 内存空间分配(四)

(5) 执行第 13 行,用对象 v 和第 10 行定义的局部变量 i 作为参数调用 another 方法,内存空间如图 3-10 所示。

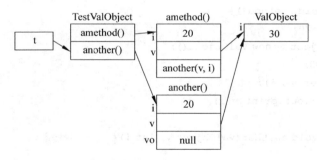

图 3-10 内存空间分配(五)

(6) 执行第 17 行,将变量 i 的值改为 0,内存空间如图 3-11 所示。
(7) 执行第 18 行,将对象 t 的变量 i 改为 40,内存空间如图 3-12 所示。
(8) 执行第 19 行,创建 ValObject 对象,引用名为 vo,内存空间如图 3-13 所示。
(9) 执行第 20 行,将 another 方法中的对象 vo 赋值给 v,使得 another 方法中的对象 v 引用 vo 的内存空间,如图 3-14 所示。
(10) 执行第 20 行,打印 v.i 的值,即打印 another 方法中对象 v 所指的变量 i 的值,从

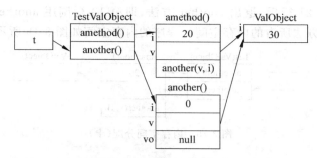

图 3-11　内存空间分配（六）

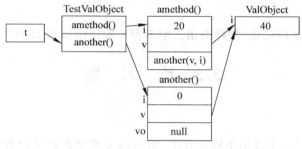

图 3-12　内存空间分配（七）

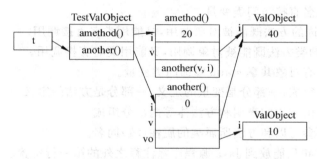

图 3-13　内存空间分配（八）

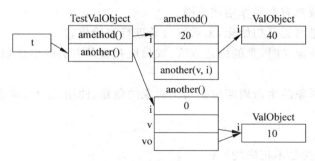

图 3-14　内存空间分配（九）

图 3-14 可知，v 对象中变量 i 的值为 10。所以，第 20 行输出 10。

（11）执行第 21 行，打印 another 方法中变量 i 的值。从图 3-14 可知，变量 i 的值为 0，所以，第 21 行输出 0。

（12）执行完第 21 行后，退出 another 方法，即第 13 行调用 another(v, i)方法结束，JVM 释放 another 方法所占的内存空间。程序的内存空间如图 3-15 所示。

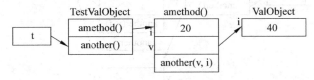

图 3-15　内存空间分配（十）

（13）执行第 14 行，打印 amethod 方法中对象 v 所指的变量 i 的值，从图 3-15 可知，对象 v 所指的变量 i 的值为 40。所以，第 14 行输出 40，方法 amethod 执行结束，即第 7 行执行结束。

（14）main 方法执行结束，程序退出。整个程序输出结果为 10040。

3.3　练　习　题

3.3.1　判断题

1. 一个类定义了一个或多个构造方法，则 Java 不提供默认的构造方法。　　　（　）
2. 如果定义的类中没有给出构造方法，系统也不会提供构造方法。　　　　　　（　）
3. 可以通过类名直接访问类变量。　　　　　　　　　　　　　　　　　　　　（　）
4. 非 static 修饰的方法既能被对象调用，又能用类名直接调用。　　　　　　　（　）
5. static 修饰的类方法既能被对象调用，又能用类名直接调用。　　　　　　　（　）
6. 一个类的所有对象共享 static 修饰的类变量。　　　　　　　　　　　　　　（　）
7. 类由两部分组成，一部分是变量的定义，一部分是方法的定义。　　　　　　（　）
8. 方法的定义由方法的声明和方法体等两部分组成。　　　　　　　　　　　　（　）
9. 方法中的形参可以和方法所属类的属性（域）同名。　　　　　　　　　　　（　）
10. package 语句只能放到 Java 源程序除注释之外的第一行位置。　　　　　　（　）
11. Java 应用程序的入口 main 方法只有一种定义。　　　　　　　　　　　　　（　）
12. 静态成员域具有全局作用域范围。　　　　　　　　　　　　　　　　　　　（　）
13. 类成员变量可无须初始化，由系统自动进行初始化。　　　　　　　　　　　（　）
14. 在类体部分定义时，类的构造方法、成员域和成员方法的出现顺序在语法上有严格限制。　　　　　　　　　　　　　　　　　　　　　　　　　　　　　　　　　　（　）
15. 类的实例对象的生命周期包括实例对象的创建、使用、废弃、垃圾的回收。（　）

3.3.2　选择题

1. 下述哪个说法是不正确的？（　　）
　A. 局部变量在使用之前无须初始化，因为有该变量类型的默认值
　B. 类成员变量由系统自动进行初始化
　C. 参数的作用域就是所在的方法
　D. for 语句中定义的变量，当 for 语句执行完时，该变量就消亡了

2. 下述哪一个关键字不是类及类成员的访问控制符？（　　）
   A. public　　　　　B. private　　　　　C. static　　　　　D. protected
3. 考虑以下的方法头声明：

   public void myMethod(int num, char letter)

   下列哪个描述是正确的？（　　）
   A. 这个方法有个 int 型的参数，它的形式参数是 letter
   B. 这个方法有个 int 型的参数，它的形式参数是 num
   C. 这个方法有个 char 型的参数，它的实际参数是 myMethod
   D. 这个方法是一个构造方法
4. 假设有个类已经定义了下述的变量：

   int num1=10;
   int num2=20;
   String word=new String("exam");
   double answer=25.5;

   另外定义了下列方法头：

   public int stuff(String s, int x, int y)

   以下哪个是正确的方法调用？（　　）
   A. num1＝stuff("hello",num1,num1);
   B. answer＝stuff(word,answer,num2);
   C. num2＝stuff("hello",num1,answer);
   D. num1＝stuff(num1,num2,word);
5. 下面哪一项不是构造方法的特点？（　　）
   A. 构造方法名必须与类名相同
   B. 构造方法不具有任何返回类型
   C. 任何一个类都含有构造方法
   D. 构造方法修饰词只能是 public
6. 以下代码的运行结果是（　　）。

```
class Book{
    public int id;                        //书的编号
    public Book( int i ){
        id=i;
    }
    protected void finalize(){
        switch (id){
        case 1:
            System.out.print( "《飘》" );
            break;
        case 2:
            System.out.print( "《Java 程序设计教程》" );
```

```
            break;
        case 3:
            System.out.print("《罗马假日》");
            break;
        default:
            System.out.print("未知书籍");
            break;
        } //switch 语句结束
        System.out.println("所对应的实例对象存储单元被回收");
    } //方法 finalize 结束
} //类 Book 结束
public class Finalize{
    public static void main(String args[ ]){
        Book book=new Book(1);
        new Book(3);
        System.gc();                    //申请立即回收垃圾
    } //方法 main 结束
}
```

A. 《罗马假日》所对应的实例对象存储单元被回收

B. 《飘》所对应的实例对象存储单元被回收

C. 《Java 程序设计教程》所对应的实例对象存储单元被回收

D. 未知书籍所对应的实例对象存储单元被回收

7. final 关键字不可以用来修饰(   )。

　　A. 类　　　　　　B. 成员方法　　　　C. 成员变量　　　　D. 接口

8. 假设已声明了一个类 A 的两个对象 a1、a2,为了初始化 a1 和 a2,下面语句正确的是(   )。

　　A. a1,a2＝new A();　　　　　　B. a1＝A.new();a2＝A.new();

　　C. a1＝new();a2＝new();　　　　D. a1＝new A();a2＝new A();

9. 以下程序的运行结果为(   )。

```
public class Test {
    int value;
    public static void main(String args[]){
        Test t=new Test();
        if(t==null){
            System.out.println("No Object");
        }
        else{
            System.out.println(t.value);
        }
    }
}
```

　　A. 0　　　　　　B. null　　　　　　C. NoObject　　　　D. 编译出错

10. 给定如下代码：

```
class Circle{
    String name, radius;
    int r;
    public Circle (String n){
        name=n;
    }
    public Circle (String n,int a){
        name=n;
        r=a;
    }
    public Circle (String n,String d,int a){
        _____
        radius=d;
    }
}
```

要在第三个构造方法的空白处填上一行代码使之能够调用第二个构造方法，这行代码应该是（   ）。

A. Circle (n,a);　　　　　　　　B. this(Circle (n,a));

C. this(n,a);　　　　　　　　　D. this(name,a);

11. 下面哪个代码片段是错误的？（   ）

A. package mypackage;
   public class Rectangle {//doing something…}

B. import java.io.*;
   package mypackage;
   public class Rectangle {//doing something…}

C. import java.io.*;
   class Circle{//doing something…}
   public class Rectangle {//doing something…}

D. import java.io.*;
   import java.awt.*;
   public class Rectangle {//doing something…}

12. 下列选项中，用于声明包名的关键字是（   ）。

A. import　　　B. package　　　C. interface　　　D. protected

13. 下列各种Java类的定义，哪些是错误的？（   ）

A. class MyClass{
       private int Val;
       public int getVal(){
           return Val;
       }
```
```

```
        }
    B. class MyClass{
           private static int Val;
           public int getVal(){
               return val;
           }
       }
    C. class MyClass{
           private int Val;
           public static int getVal(){
               return Val;
           }
       }
    D. class MyClass{
           private static int Val;
           public static int getVal(){
               return Val;
           }
       }
```

14. 某类的成员方法只能被自己调用,修饰该方法的是( )。
    A. public   B. protected   C. private   D. 无修饰符

15. 另外的类要获取以下 MyClass 类中的 member 的值,采用的语句是( )。
```
    public class MyClass{
        private static int member=1;
        public static int getMember(){
            return member;
        }
    };
```
    A. MyClass.member   B. new MyClass().member
    C. MyClass.getMember()   D. MyClass().getMember();

16. 下面关于变量的说法,哪个是不正确的?( )
    A. 实例变量是类的成员变量
    B. 实例变量用关键字 static 声明
    C. 在方法中定义的局部变量在该方法被执行时创建
    D. 局部变量在使用前必须被初始化

17. 在 Java 中,以下哪一个关键字使类不能派生出子类?( )
    A. final   B. public   C. private   D. volatile

18. 关于 public static void main 方法的参数描述不正确的是( )。
    A. String args[]   B. String[] args   C. String a[]   D. String args

19. 下列整型的最终属性 i 的定义中,正确的是(　　)。
    A. final i;   B. static int i;
    C. static final int i=234;   D. final float i=3.14f;
20. 类方法的修饰符是(　　)。
    A. static   B. class   C. public   D. void
21. 关于下面的程序,哪个说法是正确的?(　　)

    ```
    public class Person{
      static String a[]=new String[5];
      public static void main(String argv[]) {
        System.out.println(a[0]);
      }
    }
    ```

    A. 编译时将产生错误   B. 编译时正确,运行时将产生异常
    C. 输出为 null   D. 输出 0
22. 类 Line 的定义如下:

    ```
    class Line{
      private int a,b;
      public Line (int x,int y){
        this.x=x;
        this.y=y;
      }
    }
    ```

    其中的 this 代表(　　)。
    A. 类名 Line   B. 父类的对象
    C. Line 类实例的当前对象引用   D. this 指针
23. 以下程序片段,下列哪个选项插入到第 2 行将引起编译错误?(　　)

    ```
    1.   package mypackage;
    2.
    3.   class MyClass{
    4.       //do something
    5.   }
    ```

    A. public class MainClass{ }   B. package mypackage1;
    C. class B{ }   D. import java.util.*;
24. 以下程序代码,无法通过编译的是(　　)。
    A. class A{
           int i=0;
           public void method(){
             System.out.println (i);
           }

```
    }
B.  class A{
       int i=0;
    }
    class B{
       public void method(){
          System.out.println (A.i);
       }
    }
C.  class A{
       public int i=0;
       public static void method(){
          System.out.println (new A().i);
       }
    }
D.  class A{
       public static int i=0;
       public static void method(){
          System.out.println (i);
       }
    }
```

25. 下列代码在编译时会发生错误,下面哪种修改可以更正错误?(   )

```
class Test {
  private int m;
  public static void fun(){
     System.out.println(m);
  }
}
```

A. 将 private int m 改为 protected int m

B. 将 private int m 改为 public int m

C. 将 private int m 改为 static int m

D. 将 private int m 改为 int m

3.3.3　程序阅读题

1. 阅读下面的程序,程序保存为 Test.java:

```
1. public class Test{
2.    short mValue;
3.    public static void main(String[] args){
4.       int a=32;
```

```
5.        int b=56;
6.        Test os=new Test(a+b);
7.        os.Show();
8.     }
9.     protected Test(short aValue) {
10.        mValue=aValue;
11.    }
12.    public void Show() {
13.        System.out.println(mValue);
14.    }
15. }
```

上面的程序编译是否成功？如果编译出错，指出哪行出错，并说明理由；如果编译正确，运行结果是什么？

2. 阅读下面的程序：

```
1.  public class Test{
2.     public static void main(String argv[]){
3.        Bird b=new Bird();
4.        b.fly(3);
5.     }
6.  }
7.  class Bird{
8.     static int Type=2;
9.     private void fly(int an_Type){
10.       Type=an_Type;
11.       System.out.println("Flying…"+Type);
12.    }
13. }
```

上面的程序编译是否成功？如果编译出错，指出哪行出错，并说明理由；如果编译正确，运行结果是什么？

3. 仔细阅读下面的程序代码，若经编译和运行后，请写出打印结果。

```
class StaticTest {
    static int x=1;
    int y;
    StaticTest() {
    y++;
    }
    public static void main(String args[]) {
    StaticTest st=new StaticTest();
    System.out.println("x=" +x);
    System.out.println("st.y=" +st.y);
    st=new StaticTest();
    System.out.println("x=" +x);
```

```
    System.out.println("st.y="+st.y);
  }
  static {
    x++;
  }
}
```

4. 写出下列程序代码的运行结果。

```java
public class PassTest{
    float ptValue;
    public void changeInt(int value){
        value=11;
    }
    public void changeStr(String value){
        value=new String("hey");
    }
    public void changeObjValue( PassTest ref){
        ref.ptValue=22;
    }
    public static void  main(String args[ ] ){
        String str;
        int val;
        PassTest pt=new PassTest();
        val=33;
        pt.changeInt(val);
        System.out.println("Int value is: " +val);
        str=new String("Hi");
        pt.changeStr(str);
        System.out.println("Str value is: " +str);
        pt.ptValue=44;
        pt.changeObjValue(pt);
        System.out.println("Pt value is: " +pt.ptValue);
    }
}
```

5. 写出下列程序代码的运行结果：

```java
class Test {
    public static String ss="杭州电子科技大学";
    public String ms="计算机学院";
    public void printInfo(){
        System.out.println("ss="+ss+" ms="+ms);
    }
    public static void main(String [] args){
        Test obj1,obj2;
        obj1=new Test();
```

```
        obj2=new Test();
        obj1.ss="浙江工业大学";
        obj1.ms="软件学院";
        obj1.printInfo();
        obj2.printInfo();
    }
}
```

6. 阅读以下程序,回答问题。

```
package package1;
class ClassA{
    public void printInfo(){
        System.out.println("I am ClassA.");
    }
}
package package2;
import package1.ClassA;
public class ClassB{
    public static void main(String [] args){
        new ClassA().printInfo();
    }
}
```

(1) 上面两个类是在同一个源程序文件中吗?
(2) 以上程序编译是否正常?若编译正常则运行结果是什么?

7. 阅读下面程序,修改程序中错误的地方(提示:共三处)。

```
1.   class Person{
2.     String name;
3.     int age;
4.     String sex;
5.     public Person(String name,int age,String sex){
6.       this.name=name;
7.       this.age=age;
8.       this.sex=sex;
9.     }
10.    public void toString(){
11.      return "name:"+name+" age:"+age+" sex:"+sex;
12.    }
13.  }
14.  public class TestPerson{
15.    public static void main(String args[]){
16.      Person p=new Person();
17.      p.name="张三";
18.      p.age=20;
19.      p.sex="男";
```

```
20.        System.out.println(Person.tostring());
21.    }
22. }
```

8. 写出下列程序代码的运行结果。

```
public class Test{
  int x=4;
  public static void main(String argv[]){
    Test t=new Test();
    t.x++;
    change(t);
    System.out.println(t.x);
  }
  static void change(Test m){
    m.x+=2;
  }
}
```

3.3.4 编程题

1. 编写一个 Java 程序 Suansu.java,定义两个整型变量 a 和 b,使用构造方法初始化 a 为 10,b 为 5,并定义求出 a 与 b 的和(方法名为 AddAB)、差(方法名为 SubAB)、积(方法名为 MultiAB)、商（方法名为 DivAB)的方法。用另一个 Java 程序 TestSuansu.java 测试 Suansu.java 定义的方法,并在屏幕上输出结果。

2. 创建一个名为 Rectangle 的类来表示一个使用宽度和高度来改变量的矩形,矩形的宽度和高度由构造方法来确定。为 Rectangle 类创建下列方法：

getArea 返回矩形的面积,要求长和高的范围为 0~50;
getPerimeter 返回矩形的周长;
draw 使用星号(*)作为描绘字符画出该矩形(假设宽度和高度为整数);

在另一个类 TestRectangle 中编写 main 方法来测试 Rectangle 类。

3. 根据以下的设计要求编写 Java 源代码。

类名：Student
变量(访问权限均为 private)：
  name,//表示为姓名,类型为 String
  age,//表示年龄,类型为 int

方法如下。

(1) 构造方法(没有参数,设置姓名为"无名氏",年龄为 20)。
(2) setName(有一个名为 name 的 String 型参数,将 name 的值设为这个新值)。
(3) getName(没有参数,返回姓名)。
(4) setAge(有一个名为 age 的 int 型参数,将 age 的值设为这个新值)。
(5) getAge(没有参数,返回年龄)。

（6）isSameAge(有一个参数 s,是另一个 Student 对象的引用,如果两个 Student 对象的 age 相同,则返回 true,否则返回 false)。

4. 编写一个复数类 Complex 验证两个复数 1+2i 和 3+4i 相加产生一个新的复数 4+6i。复数类 Complex 必须满足如下要求。

（1）复数类 Complex 的属性如下。

realPart 是 int 型,代表复数的实数部分。

imaginPart 是 int 型,代表复数的虚数部分。

（2）复数类 Complex 的方法如下。

Complex()构造方法,将复数的实部和虚数都置 0。

Complex(int r,int i)构造方法,形参 r 为实部的初值,i 为虚部的初值。

Complex complexADD(Complex a)将当前复数对象与形参复数对象相加,所得的结果仍是一个复数值,将其返回给此方法的调用者。

String toString()把当前复数对象的实部、虚部组合成 a+bi 的字符串形式,其中,a 和 b 分别为实部和虚部的数据。

5. 编写一个表示二维平面上的点的类 MyPoint,满足以下条件。

（1）定义 private 的成员变量 x 和 y,表示点的 x 和 y 坐标,类型为整数。

（2）定义两个 MyPoint 的构造方法,一个构造方法不带参数,且 x 和 y 的初始值为 0,另一个构造方法有两个参数,参数名为 x 和 y,类型为整数,用这两个参数分别作为初始 x 和 y 的坐标。

（3）定义一个 getD 方法,功能为返回两个坐标点(由 MyPoint 定义)距离,值为 float 类型。

（4）编写 main 方法,打印坐标点(2,3)到点(4,5)的距离。

6. 编写一个公共(public)类,类名为 AccountUtil,该类属于 account 包,类中包含一个公用静态方法 toSeparateNumber,该方法以一个十进制字符串为参数,返回用逗号隔开的数字字符串,分隔从右边开始,每三个数字用一个逗号隔开。例如,若参数为 2367548,则返回 2,367,548。若参数为小数,则分隔从小数点开始,例如,若参数为 2367548.85,则返回 2,367,548.85。

根据以下说明,编写一个银行账户类 Account,并编写一个 main 方法,对 Account 类进行测试,在 main 方法中要调用 Account 的所有方法,发生异常时,要打印异常信息。该类的成员变量如表 3-1 所示(访问权限均为 private)。

表 3-1 类成员变量

变量名	含义	数据类型
Id	账号	String
owner	账户持有人姓名	String
balance	余额	double

该类的成员方法如表 3-2 所示(访问权限均为 public)。

表 3-2 类的成员方法

方法名	参 数	说 明
构造方法	无	构造一个账户实例,将 id、owner 设为 null,balance 设为 0.00
构造方法	String id, String owner, double amount	构造一个账户实例,用参数设置成员变量 id、owner、balance 的值
setID	String id	用参数设置成员变量 id 的值。返回类型 void
setOwner	String owner	用参数设置成员变量 owner 的值。返回类型 void
deposit	double amount	将金额 amount 存入账户,如果账号为 null,则抛出异常,异常信息为"账号未知!"。返回类型 double,返回值为 amount
withdraw	double amount	从账户支取金额 amount,如果账号为 null,或者余额小于 amount,则抛出异常,异常信息分别为"账号未知!"和"余额不足!"。返回类型 double,返回值为 amount
query	无	打印 id、owner、balance。返回类型 void

7. 编写两个类,classA 属于包 package1,classA 中有一个方法 methodA();classB 属于包 package2,在 classB 的方法 methodB()中调用 classA 的 methodA()方法。每个方法简单地输出方法名即可。

8. 编写一个三角形类,能根据 3 个实数构造三角形对象,如果 3 个实数不满足三角形的条件,则自动构造以最小值为边的等边三角形。输入任意 3 个数,输出构造的三角形面积。

第4章 类的封装性、继承性、多态性及接口

4.1 知 识 点

(1) 封装性，类的访问控制方式，类成员的访问控制方式，封装性的设计原则。
(2) 继承性。
(3) 多态性，包括方法重载和方法覆盖。
(4) 抽象类。
(5) 接口的定义，接口的实现，接口的作用，接口与抽象类的区别。
(6) 特殊的类，包括实名内部类、匿名内部类、泛型和Class类。
重点：类的封装性、继承性和多态性，方法的重载与覆盖、抽象类、接口、泛型。
难点：方法的重载与覆盖。

4.2 例 题

【例 4-1】 子类对象的创建与实例化过程的例子。

```
1.   class Meal {
2.      Meal() {
3.         System.out.println("Meal()");
4.      }
5.   }
6.   class Bread {
7.      Bread() {
8.         System.out.println("Bread()");
9.      }
10.  }
11.  class Cheese {
12.     Cheese() {
13.        System.out.println("Cheese()");
14.     }
15.  }
16.  class Lettuce {
17.     Lettuce() {
18.        System.out.println("Lettuce()");
19.     }
20.  }
21.  class Lunch extends Meal {
22.     Lunch() {
23.        System.out.println("Lunch()");
```

```
24.     }
25.   }
26.   class PortableLunch extends Lunch {
27.     PortableLunch(){
28.         System.out.println("PortableLunch()");
29.     }
30.   }
31.   class Sandwich extends PortableLunch {
32.     Bread b=new Bread();
33.     Cheese c=new Cheese();
34.     Lettuce l=new Lettuce();
35.     Sandwich(){
36.         System.out.println("Sandwich()");
37.     }
38.     public static void main(String[] args) {
39.         new Sandwich();
40.     }
41.   }
```

【解析】 子类对象的创建与实例化过程如下：首先，分配对象所需的全部内存空间，并初始化为 0；然后，按继承关系，自顶向下显式初始化；最后，按继承关系，从顶向下调用构造方法。因此，以上程序输出的结果如下：

```
Meal()
Lunch()
PortableLunch()
Bread()
Cheese()
Lettuce()
Sandwich()
```

【例 4-2】 下述哪一组方法，是一个类中方法重载的正确写法？

A. int addValue(int a, int b){return a+b;}
   float addValue (float a, float b) {return a+b;}

B. int addValue (int a, int b){value=a+b;}
   float addValue (int a, int b) {return (float)(a+b);}

C. int addValue(int a, int b){return a+1;}
   int addValue (int a, int b) {return a+b;}

D. int addValue(int a, int b) {return a+b;}
   int addValue (int x, int y) {return x+y;}

【解析】 一个类中，有若干个方法名字相同，但方法的参数形式不同，称为方法的重载。方法重载的原则：①方法名相同；②方法的参数类型不同，或参数个数不同，或当方法有两个以上参数时，只要参数类型的顺序不同；③与方法的参数名、返回类型和修饰符无关。所以，本例选 A。

【例 4-3】 对于下列代码：

```
public class SupperClass {
    public int sum( int a, int b) {
        return a+b;
    }
}
class SubClass extends SupperClass {   }
```

下述哪个方法可以加入类 SubClass？（   ）

A. int sum (int a, int b){ return a＋b;}
B. public void sum (int a, int b) { return;}
C. public float sum (int a, int b) { return a＋b;}
D. public int sum (int a, int b) { return a＋b;}

【解析】 从四个选项看，在子类 SubClass 加入的方法名都是 sum 且参数相同，需要与父类 SupperClass 中的 sum 方法构成覆盖关系才行。在 Java 程序中，子类和超类中有同名且参数也相同的方法，称为子类中的方法覆盖超类的方法，简称方法覆盖。子类的方法能否覆盖父类的方法，需遵守以下原则。

(1) 子类方法的名称、参数签名和返回类型必须与其父类的方法的名称、参数签名和返回类型一致，这里的参数签名指参数类型、参数个数，参数顺序都一致。

(2) 子类方法不能缩小父类方法的访问权限。比如以下程序会出现编译出错。

```
public class Base{
    public void method(){…}
}
public class Sub extends Base{
    protected   void method(){…}
                    //此行会出现编译错误，因为子类方法缩小了父类方法的访问权限
}
```

(3) 子类方法不能抛出比父类方法更多的异常。

(4) 方法覆盖只存在于子类和父类（包括直接父类和间接父类）之间。在同一个类中方法只能被重载，不能被覆盖。

(5) 父类的静态方法不能被子类覆盖为非静态的方法，反之亦然。

(6) 子类可以定义与父类的静态方法同名的静态方法，以便在子类中隐藏父类的静态方法。

(7) 父类的私有方法不能被覆盖。

(8) 父类的抽象方法可以被子类覆盖：子类实现父类的方法或重新声明父类的抽象方法。

所以，本例选 D，与父类的方法 sum 构成方法覆盖，可以加入子类。选项 A 的方法权限缩小了父类方法的访问权限，选项 B 和 C 的方法返回类型与父类方法不同。

【例 4-4】 对于下列代码：

```
public class SupperClass {
```

```
        public int sum( int a, int b) {
            return a+b;
        }
}
class SubClass extends SupperClass {   }
```

下述哪个方法不可以加入类 SubClass？（   ）

A. int sub (int a, int b){ return a-b;}

B. public int sum (int a, int b,int c){ return a+b+c;}

C. public float sum (int a, int b){ return a+b;}

D. public int sum (int a, int b) { return a+b;}

【解析】 选项 A 的方法 sub 在父类中没有，说明是子类扩展的方法，可以加入子类；选项 B 的方法 sum 有三个参数，而父类的方法 sum 只有两个参数，说明子类 SubClass 可以继承父类 SupperClass 中含有两个参数的方法，同时扩展了一个三个参数的 sum 方法，构成方法重载；选项 C 的方法名和参数与父类中 sum 方法定义都相同，但返回类型不同，不能构成方法覆盖；选项 D 的 sum 方法与父类的方法 sum 构成方法覆盖，可以加入子类。因此，本例选 C。

提示：当子类继承某一父类时，在子类中可以扩展新的方法，也可以覆盖父类的方法，也可以增加若干个方法与从父类继承来的方法构成重载关系。

【例 4-5】 继承关系中对成员访问的例子。

```
1.    class A{
2.        int x=1234;
3.        String a="aaa";
4.        void show(){
5.            System.out.println("class A : ");
6.        }
7.    }
8.    class B extends A{
9.        double x=567.89;
10.       void show(){
11.           super.show();
12.           System.out.println("class B : ");
13.            System.out.println(a);
14.           a="bbb";
15.            System.out.println(super.a);
16.            System.out.println(this.a);
17.       }
18.   }
19.   class C extends B{
20.       char x='c';
21.       void showABC(){
22.            System.out.println(super.x);
23.            System.out.println(x);
```

```
24.         System.out.println(a);
25.         super.show();
26.         System.out.println(a);
27.         show();
28.     }
29.     void show(){
30.         System.out.println("class C : ");
31.     }
32. }
33. class OverTest{
34.     public static void main(String arg[ ]){
35.         C  cc=new C();
36.         cc.showABC();
37.     }
38. }
```

【解析】 继承关系中对成员访问遵守就近原则。

（1）在子类中访问属性和方法时将优先查找自己定义的属性和方法。如果该成员在本类存在，则使用本类的，否则，按照继承层次的顺序到其祖先类查找。

（2）this 关键字特指本类的对象引用，使用 this 访问成员则首先在本类中查找，如果没有，则到父类逐层向上找。

（3）super 特指访问父类的成员，使用 super 首先到直接父类查找匹配成员，如果未找到，再逐层向上到祖先类查找。

程序输出结果如下：

```
567.89
c
aaa
class A:
class B:
aaa
bbb
bbb
bbb
class C:
```

【例 4-6】 子类对象赋值给父类对象的例子。

```
1. class Supclass{
2.     protected String className="父类属性";
3.     public void print(){
4.         System.out.println("this is 父类 print()方法"+"此时对象"+this.toString());
5.     }
6. }
7. public class Subclass extends Supclass {
8.      protected String className="子类属性";
```

```
9.      protected String otherClassName="其他子类属性";
10.     public void print(){
11.       System.out.println("this is 子类 print()方法"+"此时对象"+this.toString());
12.     }
13.     public void println(){
14.       System.out.println("this is 子类 println()方法"+"此时对象"+this.toString());
15.     }
16.     public static void main(String[] args){
17.       Supclass sup=new Subclass();
18.       System.out.println("此时的属性是:"+sup.className);
19.       sup.print();
20.       System.out.println("此时对象"+sup.toString());
21.       //此行加入 sup.println();会编译错误
22.     }
23. }
```

【解析】 将子类对象赋值给父类对象,所得到的对象是这样的一个对象:它是一个编译时为父类对象,但运行时却是一个子类对象,具体特征如下。

(1) 被声明为父类对象。
(2) 拥有父类属性。
(3) 占用子类的内存空间。
(4) 子类方法覆盖父类的方法时,此时对象调用的是子类的方法;否则,自动调用继承父类的方法。
(5) 无法访问子类中非覆盖的变量和方法。

如果在 21 行加入"sup.println();",则此行编译错误,因为在父类中没有定义 println()。如果在 21 行加入"System.out.println("此时的属性是:"+sup.otherClassName);",则此行编译也出错,Java 编译器会提示找不到符号 sup.otherClassName。

因此,本程序输出结果如下:

此时的属性是:父类属性
this is 子类 print()方法此时对象 Subclass@a90653
此时对象 Subclass@a90653

【例 4-7】 仔细阅读下面的程序代码,若经编译和运行后,请写出打印结果。

```
1.  class Man{
2.    void drink(){
3.      System.out.println("I am drinking water!");
4.    }
5.  }
6.  class OldMan extends Man{
7.    void drink(){
8.      System.out.println("I am drinking tea!");
9.    }
10. }
```

```
11.  class YoungMan extends Man{
12.    void drink(){
13.      System.out.println("I am drinking beer!");
14.    }
15.    void dance(){
16.      System.out.println("I can dance!");
17.    }
18.    public static void main(String[] args) {
19.      Man tom=new Man();
20.      Man jack=new YoungMan();
21.      Man mary=new OldMan();
22.      tom.drink();
23.      jack.drink();
24.      mary.drink();
25.      if(jack instanceof YoungMan)
26.        ((YoungMan)jack).dance();
27.    }
28.  }
```

【解析】 第25行的instanceof是关键字,它的作用是测试它左边的对象是否是它右边的类的实例,返回boolean类型的数据,这里是判断jack是否为YoungMan的实例。程序输出结果如下:

```
I am drinking water!
I am drinking beer!
I am drinking tea!
I can dance!
```

【例4-8】 抽象类的演示例子。

```
1.  abstract class Shape{
2.    abstract protected double area();
3.    abstract protected void draw();
4.  }
5.  class Rectangle extends Shape{
6.    float width,length;
7.    Rectangle(float w,float l){
8.      width=w;
9.      length=l;
10.   }
11.   public double area(){
12.     return width * length;
13.   }
14.   public void draw(){ };
15. }
16. public class ShapeDemo{
17.   public static void main(String args[]){
```

```
18.        Rectangle r=new Rectangle(6,12);
19.        System.out.println("The area of rectangle:"+r.area());
20.    }
21. }
```

【解析】 用 abstract 修饰的类为抽象类,主要用于被子类继承。设计抽象类时,应尽可能多地拥有具体类公共的代码,这样做可以提高代码的重用性;应尽可能少地拥有数据,数据应尽量放在具体类中。抽象类中一般有抽象的方法,该方法也可用 abstract 修饰,但不用实现代码,而是在子类中实现所有的抽象方法。除抽象方法外,抽象类也可有普通的成员变量或方法,也可有构造方法。由于抽象类含有抽象方法,因此抽象类不能用来生成实例,一般可通过定义子类进行实例化。可以生成抽象类的变量,该变量可以指向具体的一个子类的实例。比如有如下代码:

```
abstract class Employee{
    abstract void raiseSalary(int i);
}
class Manager extends Employee{
    void raiseSalary(int i ){ … }
}
```

可以生成抽象类 Employee 的变量 e,语句写成

```
Employee e=new Manager();
```

【例 4-9】 定义一个接口 Area,其中包含一个计算面积的方法 CalsulateArea(),然后设计 MyCircle 和 MyRectangle 两个类都实现这个接口中的方法 CalsulateArea(),分别计算圆和矩形的面积,最后写出测试以上类和方法的程序 TestArea.java。

【解析】 接口定义了一套行为规范,一个类实现这个接口就要遵守接口中定义的规范,实际上就是要实现接口中定义的所有方法。在 Java 中,接口的定义格式为[public] interface 接口名 [extends 父接口名列表],接口中仅有常量或抽象的方法,接口具有继承性,一个接口可以继承多个父接口,实现接口的格式为 class <类名> implements 接口名 1,接口名 2,…,其中一个类可以实现多个接口,类中实现接口的方法要加 public 修饰,当一个类实现了一个接口,它必须实现接口中所有的方法,这些方法都要被修饰为 public,否则会产生访问权限错误。

本例的参考代码如下:

```
1. interface Area{
2.     public double CalsulateArea();
3. }
4. class  MyCircle implements Area{
5.     double r;
6.     public MyCircle(double r){
7.         this.r=r;
8.     }
9.     public double CalsulateArea(){
10.        return Math.PI * r * r;
```

```
11.        }
12.    }
13.    class  MyRectangle implements Area{
14.        double width,height;
15.        public MyRectangle(double w,double h){
16.            width=w;
17.            height=h;
18.        }
19.        public double CalsulateArea(){
20.            return width * height;
21.        }
22.    }
23.    class TestArea{
24.        public static void main(String []args){
25.            MyCircle c=new MyCircle(2.0);
26.            System.out.println("圆的面积:"+c.CalsulateArea());
27.            MyRectangle r=new MyRectangle(2.0,3.0);
28.            System.out.println("矩形的面积:"+r.CalsulateArea());
29.        }
30.    }
```

【例 4-10】 以下是接口 I 的定义：

```
interface I{
    void setValue(int val);
    int getValue();
}
```

以下哪段代码能通过编译？（　　）

A. ```
class A extends I{
    int value;
    void setValue(int val){value=val;}
    int getValue(){return value;}
}
```

B. ```
class B implements I{
    int value;
    void setValue(int val){value=val;}
}
```

C. ```
interface C extends I{
    void increment();
}
```

D. ```
interface D implements I{
    void increment();
}
```

【解析】 接口具有继承性，一个接口可以继承多个父接口，但不能实现接口；类可以实现多个接口,但不能继承接口;类实现接口时,需实现接口中定义的所有方法。本例选 C,接口 C 可以继承父接口 I。选项 A 中类不能继承接口 I;选项 B 中的类没有实现接口 I 中定义的方法 getValue();选项 D 接口不能实现接口 I。

【例 4-11】 接口和抽象类有什么区别?

【解析】 接口和抽象类的区别主要有以下几点。

(1) 接口用 interface 修饰,而抽象类用 abstract 修饰,它们的声明格式不同。

(2) 接口具有多重继承性,可以继承多个父接口,抽象类则不可以继承多个类。

(3) 抽象类内部可以有实现的方法,接口则没有实现的方法。

(4) 接口与实现它的类不构成类的继承体系,即接口不是类体系的一部分。因此,不相关的类也可以实现相同的接口。而抽象类是属于一个类的继承体系,并且一般位于类体系的顶层。

(5) 接口的优势是通过实现多个接口实现多重继承,能够抽象出不相关类之间的相似性。而抽象类是对相关类抽象出共同的方法,主要用于被子类继承。

(6) 创建类体系的基类时,若不定义任何变量并无须给出任何方法的完整定义,则定义为接口;必须使用方法定义或变量时,考虑用抽象类。

【例 4-12】 仔细阅读下面的程序代码,若经编译和运行后,请写出打印结果。

```
1.   public class Test{
2.     private int i;
3.     public class Inner{
4.       public void increaseSize(){
5.         i++;
6.       }
7.     }
8.     Inner inner=new Inner();
9.     public void increaseSize(){
10.      inner.increaseSize();
11.    }
12.    public static void main(String[] a){
13.      Test test=new Test();
14.      for (int j=0; j<3; j++){
15.        test.increaseSize();
16.        System.out.println("i="+test.i);
17.      }
18.    }
19.  }
```

【解析】 Java 的类可以嵌套定义,某一个类内部定义的类称为内部类。内部类的类名只用于定义它的类或语句块之内,在外部引用它时必须给出带有外部类名的完整名称,并且内部类的名字不允许与外部包类的名字相同。内部类可以是抽象类或接口,若是接口,则可以由其他内部类实现。按照内部类是否含有显示的类名,可将内部类分为实名内部类和匿名内部类。

用 static 修饰的实名内部类为静态实名内部类,没有 static 修饰的内部类,称为不具有静态属性的实名内部类,它的成员域若有静态属性,则必须有 final 属性,但不能有静态属性的方法。

创建实名内部类实例的格式如下。

(1) 创建静态实名内部类实例格式:

new 外部类名.实名内部类名(构造方法调用参数列表)

(2) 创建不具有静态属性的实名内部类格式:

外部类表达式.new 实名内部类名(构造方法调用参数列表)

访问实名内部类成员的方式有以下几种格式。

(1) 访问静态实名内部类的静态成员域格式:

外部类名.实名内部类名.静态成员域名

(2) 访问静态实名内部类的不具有静态属性的成员域格式:

表达式.成员域名

(3) 调用静态实名内部类的静态成员方法格式:

外部类名.实名内部类.静态成员方法名(成员方法调用参数列表)

(4) 调用静态实名内部类的不具有静态属性的成员方法格式:

表达式.成员方法名(成员方法调用参数列表)

(5) 访问不具有静态属性的实名内部类的静态成员域格式:

外部类名.实名内部类名.静态成员域名

(6) 访问不具有静态属性的实名内部类的不具有静态属性的成员域格式:

表达式.成员域名

(7) 访问不具有静态属性的实名内部类的不具有静态属性的成员方法格式:

表达式.成员方法名(成员方法调用参数列表)

本例第 8 行创建了实名内部类 Inner 的对象,第 10 行调用不具有静态属性的实名内部类 Inner 的不具有静态属性的成员方法 increaseSize,输出结果如下:

i=1
i=2
i=3

【例 4-13】 仔细阅读下面的程序代码,若经编译和运行后,请写出打印结果。

```
1.    class C1{
2.        public void methodA(){
3.            System.out.print("A");
4.        }
5.    }
```

```
6.    interface C2{
7.        public void methodB();
8.    }
9.    class C3 extends C1 implements C2{
10.       public void methodB(){
11.           System.out.print("B");
12.       }
13.   }
14.   class C4 <T extends C1 & C2>{
15.       public void methodD( T t ){
16.           System.out.print("D");
17.       }
18.   }
19.   public class Genericity{
20.       public static void main(String args[ ]){
21.           C4<C3> a=new C4<C3>();
22.           a.methodD( new C3() );
23.       }
24.   }
```

【解析】 泛型是 Java 程序设计语言中的一种特殊类。允许程序员在 Java 中编写代码时定义一些可变部分,那些部分在使用前必须做出指明。泛型类是引用类型,是堆对象,Java 通过给类或接口增加类型参数实现。Java 泛型的类型参数只可以代表类,不能代表个别对象。类型参数的定义主要有以下三种格式。

(1) 类型变量标识符,等价于类型参数变量标识符 extends Object。

(2) 类型变量标识符 extends 父类型,表明所定义的类型变量是其父类型的子类型,如

```
public class Add<T extends java.lang.Number>{
    ⋮
}
```

(3) 类型变量标识符 extends 父类型1 & 父类型2 & … & 父类型n。

这里各父类最多仅有1个类,其余为接口。

本例第14行定义了泛型类 C4 的类型参数为 C1&C2,说明参数必须为继承类 C1 且实现接口 C2 的类型,因此,对 C4 实例化时采用第21行语句的方法,不能用"C4<C1> a=new C4<C1>();",也不能用"C4<C2> a=new C4<C2>();"本例输出结果为 D。

4.3 练 习 题

4.3.1 判断题

1. 如果类 A 和类 B 在同一个包中,则除了私有成员外,类 A 可以访问类 B 中所有的成员。   (   )

2. 接口中的成员变量全部为常量,方法为抽象方法。 （   ）
3. 抽象类可以有构造方法,所以能直接用来生成实例。 （   ）
4. Java 的类不允许嵌套定义。 （   ）
5. 包含抽象方法的类一定是抽象类,但有 abstract 修饰的类不一定包含抽象方法。
 （   ）
6. 泛型只能用于类的定义中,不能用于接口的定义中。 （   ）
7. 用 final 修饰的类不能被继承。 （   ）
8. 接口无构造器,不能有实例,也不能定义常量。 （   ）
9. 一个具体类实现接口时,必须要实现接口中的所有方法。 （   ）
10. 类具有封装性,但可以通过类的公共接口访问类中的数据。 （   ）
11. 子类能继承或覆盖(重写)父类的方法,但不能重载父类的方法。 （   ）
12. 用 final 修饰的方法不能被子类覆盖(重写)。 （   ）
13. abstract 是抽象修饰符,可以用来修饰类、属性和方法。 （   ）
14. 父类的静态方法不能被子类覆盖为非静态的方法,反之亦然。 （   ）
15. 子类实例化时,子类的构造方法一定会先调用父类的构造方法。 （   ）
16. 用 final 修饰的方法不能被覆盖(重写),也不能有重载的方法。 （   ）
17. 接口也可以继承接口,且可以继承多个接口,体现了多重继承性。 （   ）
18. 假设类 B 继承类 A,类 C 继承类 B,则在类 C 中可用 super 访问类 A 的方法。
 （   ）
19. 类和接口都可以继承另外一个类。 （   ）
20. 抽象类中不能包含 final 修饰的方法。 （   ）

4.3.2 选择题

1. Java 实现动态多态性是通过（   ）实现的。
   A. 重载　　　　　B. 覆盖　　　　　C. 接口　　　　　D. 抽象类
2. 下列哪一种描述是正确的？（   ）
   A. 动态多态性只针对静态成员方法
   B. 动态多态性只针对非静态成员方法
   C. 动态多态性只针对静态成员域
   D. 动态多态性只针对非静态成员域
3. 下列关于重载方法哪一个是正确的描述？（   ）
   A. 重载方法的参数形式(类型、参数个数或参数顺序)必须不同
   B. 重载方法的参数名称必须不同
   C. 重载方法的返回值类型必须不同
   D. 重载方法的修饰词必须不同
4. 接口的所有成员方法都具有（   ）修饰的特性。
   A. private, final   B. public, abstract
   C. static, protected   D. static

5. Java 的封装性是通过（　　）实现的。
   A. 访问权限控制　　　　　　　　B. 设计内部类
   C. 静态域和静态方法　　　　　　D. 包

6. 下列说法哪个是正确的？（　　）
   A. 子类不能定义和父类同名同参数的方法
   B. 子类只能继承父类的方法，而不能重载
   C. 重载就是一个类中有多个同名但有不同形参（类型、参数个数或参数顺序）和方法体的方法
   D. 子类只能覆盖父类的方法，而不能重载

7. 对于下列代码：

   ```
   public class Parent {
     public int addValue( int a, int b) {
       int s;
       s=a+b;
       return s;
     }
   }
   class Child extends Parent {    }
   ```

   下列哪个方法不可以加入类 Child？（　　）
   A. public int addValue(int a, int b,int c){return a+b+c;}
   B. int addValue (int a, int b){return a+b; }
   C. public int addValue(int a){return a+1;}
   D. public int addValue(int a, int b) {return a+b+1;}

8. 对于下列代码：

   ```
   1.  class Person {
   2.    public void printValue(int i, int j) {//…}
   3.    public void printValue(int i){//…}
   4.  }
   5.  public class Teacher extends Person {
   6.    public void printValue() { //…      }
   7.    public void printValue(int i) { //…     }
   8.    public static void main(String args[ ]) {
   9.      Person t=new Teacher();
   10.       t.printValue(10);
   11.   }
   12. }
   ```

   第 10 行语句将调用哪行语句？（　　）
   A. 第 2 行　　　　B. 第 3 行　　　　C. 第 6 行　　　　D. 第 7 行

9. 以下程序段输出结果的是（　　）。

   ```
   public class A implements B {
   ```

```
        public static void main(String args[]) {
            int i;
            A c1=new A();
            i=c1.k;
            System.out.println("i="+i);
        }
    }
    interface B {
        int k=10;
    }
```

    A. i=0   B. i=10   C. 程序有编译错误   D. i=true

10. 阅读下面的程序,输出结果是( )。

```
    public class TestDemo {
        int m=5;
        public void some(int x) {
            m=x;
        }
        public static void main(String args []) {
            new Demo().some(7);
        }
    }
    class Demo extends TestDemo {
        int m=8;
        public void some(int x) {
            super.some(x);
            System.out.println(m);
        }
    }
```

    A. 5   B. 8   C. 7   D. 编译错误

11. 下述哪个方法不可以加入类 SubClass?( )

```
    class SupClass{
      public void methodOne(int i){}
      public void methodTwo(int i){}
      public static void methodThree(int i){}
      public static void methodForth(int i){}
    }
    class SubClass extends SupClass{
        ⋮
    }
```

    A. public static void methodOne(int i){ }

    B. public void methodTwo(int i){ }

    C. public static void methodThree(int i,int j){ }

D. public static void methodForth(int i){ }

12. 关于下面的程序,说法正确的是(　　)。

```
class Base{
    int m;
    public Base(int m){
        this.m=m+1;
    }
}
public class Test extends Base {
    public Test(){
        m=m+1;
    }
    public static void main(String args[]){
        Test t=new Test();
        System.out.print(t.m);
    }
}
```

A. 输出结果为 0　　　　　　　　B. 输出结果为 1
C. 输出结果为 2　　　　　　　　D. 编译出错

13. 关于下面的程序,编译和运行后输出结果是(　　)。

```
class Base{
    int m=0;
    public int getM(){
        return m;
    }
}
public class Test extends Base {
    int m=1;
    public int getM(){
        return m;
    }
    public static void main(String args[]){
        Test t=new Test();
        System.out.print(t.m);
        System.out.print(t.getM());
    }
}
```

A. 00　　　　B. 01　　　　C. 10　　　　D. 11

14. 设有下面的两个类定义:

```
class  A {
    void  Show(){
        System.out.println("我喜欢 Java!");
```

```
    }
}
class B extends A{
    void Show(){
        System.out.println("我喜欢C++!");
    }
}
```

则顺序执行如下语句后输出结果为(   )。

```
A  a=new A();
B  b=new B();
a.show();
b.show();
```

A. 我喜欢Java!
　　我喜欢C++!

B. 我喜欢C++!
　　我喜欢Java!

C. 我喜欢Java!
　　我喜欢Java!

D. 我喜欢C++!
　　我喜欢C++!

15. 现有两个类A和B,以下描述中表示B继承A的是(   )。
    A. class A extends B   B. class B implements A
    C. class A implements   D. class B extends A

16. 定义类B和类C如下,并将其保存为B.java文件,得到的结果是(   )。

```
class B{
  int b;
  B(int i){
      b=i;
  }
}
class C extends B{
    double c=7.8;
}
```

　　A. 代码能够成功编译运行
　　B. 代码无法编译因为类B不是一个应用程序或小程序
　　C. 代码无法编译,因为类C没有定义一个带参数的构造方法
　　D. 代码无法编译,因为类B没有定义一个不带参数的构造方法

17. 类Teacher和Student都是类Person的子类,t、s、p分别是上述三个类的非空引用变量,关于以下语句说法正确的(   )。

```
if(t instanceof Person){
    s=(Student)t;
}
```

　　A. 将构造一个Student对象   B. 表达式合法
　　C. 编译时非法   D. 编译时合法而在运行时可能非法

18. 在// point x 处的哪个声明是合法的？（　　）

```
class Person {
  private int a;
  public int change(int m){
    return m;
  }
}
public class Teacher extends Person {
  public int b;
  public static void main(String arg[]){
    Person p=new Person();
    Teacher t=new Teacher();
    int i;
    //point x
  }
}
```

  A．i＝m；　　　　　　　　　　　B．i＝b；

  C．i＝p.a；　　　　　　　　　　D．i＝p.change(30)；

19．下面关于继承的叙述哪些是正确的？（　　）

  A．在 Java 中的类只允许继承一个类

  B．在 Java 中一个类允许继承多个类

  C．在 Java 中一个类不能同时继承一个类和实现一个接口

  D．在 Java 中接口可以继承一个或多个接口

20．下列哪些方法与方法 public void add(int a){ }构成重载方法？（　　）

  A．public int add(int a)　　　　　B．public long add(long a)

  C．public void add(int a,int b)　　D．public void add(float a)

21．在 Java 语言中，类 Cat 是类 Animal 的子类，Cat 的构造方法中有一句 super()，该语句表达了什么含义？（　　）

  A．调用类 Cat 中定义的 super()方法

  B．调用类 Animal 中定义的 super()方法

  C．调用类 Animal 的构造方法

  D．语法错误

22．定义一个类名为 MyClass.java 的类，并且该类可被一个工程中的所有类访问，那么该类的正确声明应为（　　）。

  A．private class MyClass extends Object

  B．class MyClass extends Object

  C．public class MyClass extends Object

  D．protected class MyClass extends Object

23．关于下面的程序，以下哪个结论是正确的？（　　）

  1. public class Test {

```
2.     public Test(){
3.         System.out.print("3");
4.     }
5.     public void Test(){
6.         System.out.print("2");
7.     }
8.     public static void main(String args[]){
9.        Test t=new Test();
10.       t.Test();
11.       System.out.print("1");
12.    }
13. }
```

A. 程序可以通过编译并正常运行,输出结果为 21

B. 程序可以通过编译并正常运行,输出结果为 31

C. 程序可以通过编译并正常运行,输出结果为 321

D. 程序无法通过编译

24. 给定如下 Java 程序:

```
class A{
  public A (){
    System.out.print("A");
  }
}
class B extends A{
  public B(){
    System.out.print("B");
  }
  public static void main(String[] args){
     B b=new B();
  }
}
```

上述程序将(　　)。

A. 不能通过编译

B. 通过编译,执行后输出为 AB

C. 通过编译,执行后输出为 B

D. 通过编译,执行后输出为 A

25. 已知 MyInterface 是一个接口,ClassA 是实现了 MyInterface 的一个类,ClassB 是 ClassA 的子类,则下列哪个语句是正确的?(　　)

A. ClassB obj＝new ClassA();

B. MyInterface obj＝new ClassB();

C. ClassA obj＝new MyInterface();

D. MyInterface obj＝new MyInterface();

26. A 派生出子类 B,B 派生出子类 C,并且在 Java 源代码中有如下声明：

    A a0=new A();
    A a1=new B();
    A a2=new C();

    以下哪个说法是正确的？（   ）
    A. 只有第 1 行能通过编译
    B. 第 1 行和第 2 行能通过编译,但第 3 行编译出错
    C. 第 1~3 行能通过编译,但第 2 行和第 3 行运行时出错
    D. 第 1 行、第 2 行和第 3 行的声明都是正确的

27. 考虑以下代码：

    class C1{ }
    interface C2{ }
    class C3 extends C1 implements C2{ }
    class C4 <T extends C1 & C2>{ }

    则下列哪个语句是正确的？（   ）
    A. C4<C1> c41＝new C4<C1>();
    B. C4<C2> c42＝new C4<C2>();
    C. C4<C3> c43＝new C4<C3>();
    D. C4<T> c44＝new C4<T>();

28. 对于以下类：

    class A{ }
    class B extends A{ }
    class C extends A{ }
    public class Test {
        public static void main(String args[]) {
            A x=new A();
            B y=new B();
            C z=new C();
            //此处插入一条语句
        }
    }

    下面哪个语句可以放到插入行？（   ）
    A. x＝y;   B. z＝x;
    C. z＝(c)y;   D. y＝(a)y;

29. 设有一个类的代码如下：

    class Outer{
      public class Inner1{ }
      public static class Inner2{ }
    }

则在另一个类的代码中,下列哪个语句是正确的?(　　)

A. Outer.Inner1 obj=new Outer.Inner1();

B. Outer.Inner2 obj=new Outer.Inner2();

C. Outer.Inner1 obj=new Outer.Inner1().new Inner1();

D. Outer.Inner2 obj=new Outer().new Inner2();

30. 以下关于泛型的说法哪个是错误的?(　　)

A. 泛型是通过类型参数来提高代码复用性的一种技术

B. 通过在类名后增加类型参数可以定义具有泛型特点的类

C. 通过在接口名后增加类型参数可以定义具有泛型特点的接口

D. 一个泛型类只能有一个类型参数

4.3.3　程序阅读题

1. 仔细阅读下面的程序代码,若经编译和运行后,请写出打印结果。

```
class Overload {
    void testOverload(int i) {
        System.out.println("int");
    }
    void testOverload(String s) {
        System.out.println("String");
    }
    public   static   void main(String args[]){
        Overload  a=new  Overload ();
        char ch='x';
        a.testOverload(ch);
    }
}
```

2. 仔细阅读下面的程序代码,请将划线上(1)～(5)的语句补充完整。

```
abstract class Person{
  private String name;
  public Person(String n){
    name=n;
  }
  public   (1)             String getMajor();
  public String    (2)            (){
    return name;
  }
}
class Student    (3)            Person{
  private    (4)            ;
  public Student(String n, String m){
    super(n);
```

```
    major=m;
  }
  public String     (5)         (){
    return "专业是:" +major;
  }
}
public class TestPerson{
  public static void main(String args[]){
    Person p=new Student("张三","软件工程");
    System.out.println(p.getName() +", "+p. getMajor());
  }
}
```

3. 写出下列程序代码的运行结果。

```
public class Test {
  int m=1;
  public  void some(int x) {
    m=x;
  }
public static void main(String args []) {
    new Demo().some(2);
  }
}
class Demo extends Test{
  int m=3;
  public  void some(int x) {
    super.some(x);
    System.out.print(m);
    System.out.print(super.m);
  }
}
```

4. 写出下列程序代码的运行结果。

```
class A{
  int m=0,n=0;
  long f (){
    return m+n;
  }
}
class B extends A{
  int m=1,n=1;
  long f(){
    long result=0;
    super.m=10;
    super.n=30;
```

```java
      result=super.f()+(m+n);
      return result;
    }
    long g(){
      long result=0;
      result=super.f();
      return result/2;
    }
}
class Example{
  public static void main(String args[ ]){
    B b=new B();
    b.m=6;
    b.n=2;
    long resultOne=b.g();
    long resultTwo=b.f();
    long resultThree=b.g();
    System.out.println("resultOne="+resultOne);
    System.out.println("resultTwo="+resultTwo);
    System.out.println("resultThree="+resultThree);
  }
}
```

5. 下面的程序运行结果是什么？

```java
class Tree{}
class Pine extends Tree{}
class Oak extends Tree{}
public class Forest{
  public static void main( String[] args ) {
    Tree tree=new Pine();
    if( tree instanceof Pine )
      System.out.println( "Pine" );
    if( tree instanceof Tree )
      System.out.println( "Tree" );
    if( tree instanceof Oak )
      System.out.println( "Oak" );
    else
      System.out.println( "Oops" );
  }
}
```

6. 下面的程序运行结果是什么？

```java
abstract class Base{
  abstract public void myfunc();
  public void another(){
```

```
      System.out.println("Another method");
    }
}
public class Abs extends Base{
    public static void main(String argv[]){
      Base b=new Abs();
      b.another();
    }
    public void myfunc(){
      System.out.println("My Func");
    }
    public void another(){
      myfunc();
    }
}
```

7. 下面的程序运行结果是什么？

```
class Super{
    public int i=0;
    public Super(){
      i=1;
    }
}
public class Sub extends Super{
    public Sub(){
      i=2;
    }
    public static void main(String args[]){
      Sub s=new Sub();
      System.out.println(s.i);
    }
}
```

8. 下面的程序运行结果是什么？

```
class Person{
    public Person(){
      System.out.println("hi!");
    }
    public Person(String s){
      this();
      System.out.println("I am "+s);
    }
}
public class Who extends Person{
    public Who(){
```

```
    this("I am Tony");
  }
  public Who(String s){
    super(s);
    System.out.println("How do you do?");
  }
  public static void main(String args[]){
    Who w=new Who("Tom");
  }
}
```

9. 阅读下面的程序,修改程序中错误的地方(提示：共三处错误)。

```
1.  interface Shape{
2.    double PI;
3.    double area();
4.  }
5.  class Cycle extends Shape{
6.    private double r;
7.    public Cycle(double r){
8.      this.r=r;
9.    }
10.    double area(){
11.      return PI * r * r;
12.    }
13.  }
14. public class Test{
15.    public static void main(String args[]){
16.      Cycle c=new Cycle(1.5);
17.      System.out.println("面积为:"+c.area());
18.    }
19. }
```

10. 仔细阅读下面的程序代码,若经编译和运行后,请写出打印结果。

```
class GenericsFoo<T>{
  private T x;
  public GenericsFoo(T x) {
    this.x=x;
  }
  public T getX() {
    return x;
  }
  public void setX(T x) {
    this.x=x;
  }
}
```

```java
public class GenericsFooDemo {
  public static void main(String args[]){
    GenericsFoo<String> strFoo=new GenericsFoo<String>("Hello!");
    GenericsFoo<Double> douFoo=new GenericsFoo<Double>(new Double("1"));
    System.out.println("strFoo.getX="+strFoo.getX());
    System.out.println("douFoo.getX="+douFoo.getX());
  }
}
```

11. 阅读下面的程序代码，写出程序运行的输出结果。

```java
class ParentClass{
  int x=0;
  int sum(int a,int b, int c){
    return a+b+c;
  }
  int sum(int a,int b){
    return a+b;
  }
}
class ChildClass extends ParentClass{
  public ChildClass() {
    x=10;
  }
  int sum(int a,int b){
    return a+b+1;
  }
}
class Test{
  public static void main(String args[]){
    ParentClass p=new ChildClass();
    System.out.println(p.sum(5,5,5));
    System.out.println(p.sum(5,5));
    System.out.println(p.x);
  }
}
```

12. 仔细阅读下面的程序代码，写出程序运行的输出结果。

```java
public class TestSample {
    public static void main(String args[]) {
        Sub obj1=new Sub();
        Super obj2=new Sub();
        Super obj3=new Super();
        System.out.println(obj1.method1());
        System.out.println(obj2.method1());
        System.out.println(obj3.method1());
```

```
    }
}
class Super {
    int x=1, y=2;
    int method1() {
        return (x<y ? x : y);
    }
}
class Sub extends Super {
    int method1() {
        return ((x>y) ? x : y);
    }
}
```

13. 仔细阅读下面的程序代码,写出程序运行的输出结果。

```
class Test1 {
    private int i=1;
    public class Test11{
        private int i=2;
        public void methodI(int i){
            i++;
            this.i++;
            Test1.this.i++;
            System.out.println("i of methodI():"+i);
            System.out.println("i of Test11:"+this.i);
            System.out.println("i of Test1:"+Test1.this.i);
        }
    }
    Test11 ic=new Test11();
    public void increaseI(int k){
        ic.methodI(k);
    }
    public static void main(String [] args){
        Test1 oc=new Test1();
        oc.increaseI(20);
    }
}
```

14. 阅读下面的程序代码,判断 26～35 行(带划线部分)各语句编译是否通过,如果编译通过,直接写出该行的打印结果。

```
1.  class ParentClass{
2.      int x=0;
3.      int sum(int a,int b){
4.          return a+b;
5.      }
```

```
6.      int sub(int a,int b){
7.          return a-b;
8.      }
9.  }
10. class ChildClass extends ParentClass{
11.     int x=1;
12.     int y=2;
13.     int sum(int a,int b){
14.         return a+b+1;
15.     }
16.     int multi(int a,int b){
17.         return a * b;
18.     }
19. }
20. class Test{
21.     public static void main(String args[]){
22.         ParentClass p=new ParentClass();
23.         ChildClass c=new ChildClass();
24.         ParentClass pp=c;
25.         ChildClass cc=(ChildClass)pp;
26.         System.out.println(pp.sum(5,5));
27.         System.out.println(pp.sub(10,5));
28.         System.out.println(pp.multi(10,5));
29.         System.out.println(pp.x);
30.         System.out.println(pp.y);
31.         System.out.println(cc.sum(5,5));
32.         System.out.println(cc.sub(10,5));
33.         System.out.println(cc.multi(5,5));
34.         System.out.println(cc.x);
35.         System.out.println(cc.y);
36.     }
37. }
```

4.3.4 编程题

1. 按以下要求编写程序。

（1）根据下面的要求实现圆类 Circle：

Circle 类的成员变量：radius 半径。

Circle 类的方法成员如下所示。

Circle()：构造方法，将半径置为 0。

Circle(double r)：构造方法，创建 Circle 对象时将半径初始化为 r。

double getRadius()：获得圆的半径值。

double getPerimeter()：获得圆的周长。

double gerArea()：获得圆的面积。

void disp()：将圆的半径、周长、面积输出到屏幕上。

（2）继承上题中的圆 Circle 类，派生圆柱体类 Cylinder，要求如下。

Cylinder 类的成员变量：height 表示圆柱体的高。

Cylinder 类的方法成员如下所示。

Cylinder(double r,double h)构造方法：创建 Cylinder 对象时将圆半径初始化为 r，圆柱高初始化为 h。

double getHeight()：获得圆柱体的高。

double getCylinderArea()：获得圆柱体的面积。

double getVol()：获得圆柱体的体积。

void dispVol()：将圆柱体的体积输出到屏幕。

2. 按以下要求编写程序。

（1）定义一个 Shape 接口，该接口中只有一个抽象方法 getArea()，该方法无参数，返回值类型为 double 型。

（2）定义一个圆类 Circle，满足以下条件。

① Circle 类实现 Shape 接口。

② 定义 Circle 类的成员变量 r，表示圆的半径，数据类型为 int。

③ 定义 Circle 类的构造方法，参数名为 r，用该参数初始化圆的半径。

④ 实现 getArea()方法，计算圆的面积（圆周率取 3.14）。

（3）编写一个测试类，类名为 TestCircle，利用 Circle 类计算半径为 5 的圆的面积，并将面积在屏幕打印出来。

3. 定义一个接口 Shape，其中包括一个方法 area()，设计"三角形"、"圆"、"长方形"等类实现 Shape 接口。分别创建一个"三角形"、"圆"、"长方形"对象存入一个类型为 Shape 的数组中，将数组中各类图形的面积输出。

4. 请编程实现以下要求：

（1）定义一个接口 DataStructure，包括以下方法：

```
Boolean isFull()
Boolean isEmpty()
Void addElement(Object obj)
Object removeElement()
```

（2）设计一个队列类（MyQueue）实现 DataStructure 接口。队列的大小由其构造方法指定。要求实现的方法体现出队列的先进先出特性。

（3）设计一个测试类，在其主方法中用 DataStructure 类型的引用变量引用一个大小为 10 的 MyQueue 对象，使用 addElement()方法增加"0"，"1"，…，"9"共 10 个字符串对象，再用 removeElement()方法取出这些元素并打印出来。

5. 按以下要求编程程序。

（1）编写一个抽象类 Animal，其成员变量有 name、age、weight 表示动物名、年龄和质量。方法有 showInfo()、move()和 eat()，其中后面两个方法是抽象方法。

（2）编写一个类 Bird 继承 Animal，实现相应的方法。通过构造方法给 name、age、weight 分别赋值，showInfo()打印鸟名、年龄和质量，move()方法打印鸟的运动方式，eat()

打印鸟喜欢吃的食物。

（3）编写测试类 TestAnimal，用 Animal 类型的变量，调用 Bird 对象的三个方法。

6. 尽量少写相同的代码编写程序描述如图 4-1 所示的类层次，其中人为父类 Person，其属性包括姓名、性别、出生日期等，方法为 printInfo() 打印信息。教师 Teacher 还包括学校和工号属性；学生 Student 还包括学校、学号、专业、年级和班级等属性；编写一个测试类 TestPerson，在 main 方法中创建 1 名教师和 1 名学生对象，输出对象的所有属性信息。

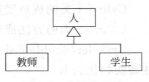

图 4-1 人的类层次

7. 为管理学校中教师的工作证和学生的学生证设计一个类体系结构，尽可能保证代码的重用率。假设教师工作证包括编号、姓名、出生年月、部门、职务和签发工作日期；学生证包括编号、姓名、出生年月、学院、专业、入校时间及每学年的注册信息等。

8. 按以下要求编写程序。

（1）定义接口 AreaInterface，该接口有一个双精度浮点型的常量 PI，它的值等于 Math.PI；含有一个求面积的方法 double area()。

（2）定义一个 Rectangle(长方形)实现 AreaInterface 接口,该类有两个 private 访问权限的双精度浮点型变量 x(长)和 y(宽)；定义一个 public 访问权限的构造方法，用来给类变量赋值；实现 area() 方法得到长方形的面积；定义 toString() 方法，返回一段字符串信息，内容如下格式："该长方形面积为："+面积。

（3）定义一个 TestArea 类，在它的 main() 方法中；创建一个 Rectangle 的实例，长为 10.0，宽为 20.0，输出它的面积。

9. 编写一个汽车类 Car。

Car 具有以下属性。

品牌：brand，类型为 String；发动机排量：engineDisplacement，类型为 double；速度：speed，类型为 double；状态：status，类型为 boolean；最高时速：maxSpeed，类型为 double。

Car 具有以下方法。

构造方法：Car(String brand, double engineDisplacement, double maxSpeed)，该方法使用参数设置成员变量的值。

启动：start()，该方法使得 status 的值变成 true。

加速：speedUp()，当汽车处于启动状态时，该方法每调用一次，速度增加 5，但速度不得高于最高时速。

减速：slowDown()，当汽车处于启动状态时，该方法每调用一次，速度减 5，但速度不得小于 0。

熄火：stop()，当 speed 为零时，将 status 的值变成 false。

每个方法除了改变成员变量的之外，还要打印出方法执行后的状态和速度。

编写 main 方法实例化一个 Car 对象，其品牌为"红旗"，排量为 2.0，最高时速为 160.00，启动该汽车，加速到 120，再减速到 0，最后熄火。

10. 运输工具 Vehicle 都有名称（name）、品牌（brand）、最大载重量（loadCapacity）、当前载重量（load）、最高速度（maxSpeed）、速度（speed）等属性，也都有移动（move）、加速（speedUp）、减速（slowDown）、停止（stop）等方法。飞机 Plane、汽车 Car、轮船 Ship、马车

Wagon 概不例外。虽然都有 move 和 stop 方法，但其实现不同。其他方法的实现相同。请编写相应的类，尽量少写相同的代码。编写一个测试类 TestVehicle，在其 main 方法中声明一个 Vehicle 类型的引用变量 vehicle，分别引用一个 Plane、Car、Ship、Wagon 对象，并执行相应的方法。

11. 在一个类 Outer 中定义了属性 name 和 i，其构造方法将 name 赋值为 Outer，i 赋值为 20，在 Outer 中定义一个内部类 Inner，也定义了属性 name 和 i，并将其初始化为 Inner 和 10；在内部类中编写一个方法 printInfo()，输出外部类和内部类中所有的属性值。

第 5 章　数组、字符串和枚举

5.1　知　识　点

(1) 数组，包括一维数组和多维数组。
(2) 字符串，包括 String 和 StringBuffer。
(3) 字符串与其他数据类型的转换。
(4) 枚举的定义、变量和常量以及使用方法。

重点：一维数组，String 和 StringBuffer，字符串与其他数据类型的转换，枚举的定义和使用方法。

难点：一维数组的应用，String 和 StringBuffer 的应用。

5.2　例　　题

【例 5-1】　编写一个程序实现如下方法：
(1) 求一个数组的最大元素值：public static int maxElement(int a[])。
(2) 求一个数组的所有元素平均值：public static int average(int a[])。
(3) 查找某个数在数组中的位置：public static int search(int a[], int b)。

利用随机数产生 20 个整数给一个数组赋值，分别验证以上方法。

【解析】　Java 中的数组是相同类型的数据元素按顺序组成的一种引用数据类型，元素在数组中的相对位置由下标来指明。Java 数组实际上也是对象，所以可通过 new 关键字来创建，如 int[ ] c=new int[12]；数组的长度用数组名.length，如 c.length。数组元素用数组名[下标]表示，下标的范围是 0～数组长度－1，如 c[0]。本例可用一个循环可实现对一维数组所有元素的遍历访问，从而实现最大值和累加值。为实现元素的查找，可根据值从前往后查找其出现位置，返回在数组中出现第一次出现位置，如果未查到，则返回－1。

本例的参考代码如下：

```java
public class TestArray {
    /*求数组的最大元素值*/
    public static int maxElement(int a[]){
        int m=a[0];
        for (int k=1;k<a.length;k++){
            if(m<a[k])
                m=a[k];
        }
        return m;
    }
    /*求所有元素的平均值*/
```

```java
    public static int average(int a[]){
        int s=0;
        for (int k=0;k<a.length;k++){
            s=s+a[k];
        }
        return s/a.length;
    }
    /*查找数 b 在数组 a 中的首次出现位置*/
    public static int search(int a[],int b){
        for (int k=0;k<a.length;k++){
            if(a[k]==b)
                return k;
        }
        return -1;
    }
    public static void main(String args[]){
        int b[]=new int[20];//定义数组、大小
        /*给数组赋值并输出数组*/
        for(int k=0;k<b.length;k++){
            b[k]=(int)(Math.random() * 20);
            System.out.print(b[k]+",");
        }
        System.out.println();
        System.out.println("数组 b 的最大值是:"+maxElement(b));
        System.out.println("数组 b 的平均值是:"+average(b));
        int x=(int)(Math.random() * 20);
        int p=search(b,x);
        if(p!=-1){
            System.out.println(x+"在数组 b 的位置为"+p);
        }
        else{
            System.out.println(x+"不在数组中!");
        }
    }
}
```

【例 5-2】 字符串 String 对象的比较。假设有以下代码：

```java
String s1="hello";
String s2="hello";
String s3=new String("hello");
```

那么,以下 1～4 个表达式语句的值分别是什么？

1. s1==s2;
2. s1==s3;
3. s1.equals(s2);

4. s1.equals(s3);

【解析】 在 Java 中,String 是一个特殊的类,可用 new 创建字符串对象,也可像基本数据类型一样创建 String,如 String s1="hello";这里的 s1 为字符串常量。JVM 有一个专门存放字符串常量的内存空间,称为字符串池。为存储 s1 的值"hello",JVM 首先从为字符串池查找是否有"hello",若有,则 s1 引用"hello"的内存空间,否则就在字符串池存放"hello"。对字符串对象来说,"=="是比较字符串对象引用的内存地址,而 equals 则比较字符串内容是否相同。本例第 1、第 3 和第 4 个表达式语句的值都为 true,而第 2 个表达式的值为 false。

【例 5-3】 枚举类型应用程序的示例。

```
enum Season{
    春季,夏季,秋季,冬季
}
public class EnumDemo{
    public static void main(String args[ ]){
        Season [ ] sa=Season.values();
        for ( int i=0; i<sa.length; i++){
            switch(sa[i]){
            case 春季:
                System.out.println("春季花满天");
                break;
            case 夏季:
                System.out.println("夏季热无边");
                break;
            case 秋季:
                System.out.println("秋季果累累");
                break;
            case 冬季:
                System.out.println("冬季雪皑皑");
                break;
            } //switch 结构结束
        } //for 循环结束
    } //方法 main 结束
} //类 EnumDemo 结束
```

【解析】 创建枚举类型的主要目的是为了定义一些枚举常量。不能通过 new 运算符创建实例对象,可直接通过枚举类型标识符访问枚举变量。如 Season s1=Season.春季;也可通过枚举变量访问枚举常量,如 s.夏季==Season.夏季。本例枚举类型可直接用于 switch…case 语句。

5.3 练 习 题

5.3.1 判断题

1. 数组是 Java 基本类型的一种。   (   )

2. 一维数组一旦创建,其大小是不能被随意改变的。   (   )
3. StringBuffer 的实例对象所包含的字符序列可以被修改。   (   )
4. 在 Java 中,"abc".length()的值是 3。   (   )
5. 数组类型不是一种基本数据类型,它是一种引用数据类型。   (   )
6. 枚举类型和类一样,可通过 new 创建实例对象。   (   )
7. 用"＋"可以实现字符串的拼接,用"－"可以从一个字符串中去除一个字符子串。
                                                      (   )
8. 枚举类型中也可以定义成员变量和方法。   (   )

5.3.2 选择题

1. 下列创建一维数组合法的语句是(　　)。
   A. int[] a＝new int[10];　　　　　B. int[] a＝new int(10);
   C. int a[];　　　　　　　　　　　　D. int[] a;
2. 返回字符串长度的是(　　)。
   A. 成员域 length　　　　　　　　　B. 静态成员域 length
   C. 实例方法 length()　　　　　　　D. 静态成员方法 length()
3. new Vector()创建的向量容量为(　　)。
   A. 16　　　　　B. 8　　　　　C. 10　　　　　D. 0
4. 关于下面的程序,哪个结论是正确的?(　　)

```
public class Test {
    public static void main(String args[]){
        int a[]=new int[5];
        a.length=10;
        System.out.println(a.length);
    }
}
```

   A. 程序可以通过编译并正常运行,结果输出 10
   B. 程序可以通过编译并正常运行,结果输出 5
   C. 程序无法通过编译
   D. 程序可以通过编译,但无法正常运行
5. 对于下列代码:

```
String str1="java";
String str2="java";
String str3=new String("java");
StringBuffer str4=new StringBuffer("java");
```

   以下表达式的值为 true 的是(　　)。
   A. str1＝＝str2;　　　　　　　　　B. str1＝＝str4;
   C. str2＝＝str3;　　　　　　　　　D. str3＝＝str4;

6. 以下程序段输出结果的是（　　）。

```
1.   public class Test {
2.     public static void main(String args[]){
3.       String str="ABCDE";
4.       str.substring(3);
5.       str.concat("XYZ");
6.       System.out.print(str);
7.     }
8.   }
```

A. DE　　　　　B. DEXYZ　　　　C. ABCDE　　　　D. CDEXYZ

7. 下列声明二维数组的语句不合法的是（　　）。

A. char c[][]=new char[2][3];　　　B. char c[][]=new char[6][];
C. char [][]c=new char[3][3];　　　D. char [][]c=new char[][4];
E. int []a[]=new int[10][10];

8. 某个main()方法中有以下代码：

```
double[] num1;
double num3=2.0;
int num2=5;
num1=new double[num2+1];
num1[num2]=num3;
```

请问以上程序编译运行后的结果是（　　）。

A. num1 指向一个有 5 个元素的 double 型数组
B. num2 指向一个有 5 个元素的 int 型数组
C. num1 数组的最后一个元素的值为 2.0
D. num1 数组的第 3 个元素的值为 5

9. 以下代码的运行结果是（　　）。

```
public class Example{
  String str=new String("hello");
  char ch[]={'d','b','c'};
  public static void main(String args[]){
      Example ex=new Example();
      ex.change(ex.str,ex.ch);
      System.out.println(ex.str+" and "+ex.ch[0]);
  }
  public void change(String str,char ch[]){
      str="world";
      ch[0]='a';
  }
}
```

A. hello and d　　　　　　　　　B. hello and a

C. world and d   D. world and a

10. 以下代码的运行结果是(　　)。

```java
public class StringTest{
  public static void operate(String x,String y){
    x.concat(y);
    y=x;
  }
  public static void main (String args[]){
    String a="A";
    String b="B";
    operate(a,b);
    System.out.println(a+"."+b);
  }
}
```

A. A.A   B. B.A   C. A.B   D. B.B

11. 以下代码的运行结果是(　　)。

```java
public class StringArrayTest{
  public static void swap(String [] s){
    if(s.length<2)
      return;
    String t=s[0];
    s[0]=s[1];
    s[1]=t;
  }
  public static void main (String args[]){
    String [] s={"1","2"};
    swap(s);
    System.out.print(s[0]+s[1]);
  }
}
```

A. 20   B. 21   C. 22   D. 23

12. 以下代码的运行结果是(　　)。

```java
public class TestGetChars{
  public static void main(String[] args){
    String myString=new String("Hello World!");
    char [] yourString=new char[5];
    myString.getChars(6,11,yourString,0);
    System.out.println(myString);
    System.out.println(yourString);
  }
}
```

A. Hello World!
   Hello
B. Hello World!
   Hello!
C. Hello World!
   World
D. World
   World

13. 以下代码的运行结果是（　　）。

```
public class TestByteArrayString{
  public static void main(String[] args){
    byte [] b={49,50,51,52,53};
    String myString=new String(b);
    System.out.println(myString);
  }
}
```

A. 45555 B. 12345 C. 90123 D. 4950515253

14. 以下代码的运行结果是（　　）。

```
1.  public class TestIntern{
2.    public static void main(String args[ ]){
3.      String s1="123456";
4.      String s2="123456";
5.      String s3="123" +"456";
6.      String a0="123";
7.      String s4=a0 +"456";
8.      String s5=new String("123456");
9.      String s6=s5.intern();
10.     System.out.println("s2" + ((s2==s1) ?"==" : "!=") +"s1");
11.     System.out.println("s3" + ((s3==s1) ?"==" : "!=") +"s1");
12.     System.out.println("s4" + ((s4==s1) ?"==" : "!=") +"s1");
13.     System.out.println("s5" + ((s5==s1) ?"==" : "!=") +"s1");
14.     System.out.println("s6" + ((s6==s1) ?"==" : "!=") +"s1");
15.   }
16. }
```

A. s2==s1
   s3==s1
   s4!=s1
   s5!=s1
   s6==s1

B. s2==s1
   s3==s1
   s4==s1
   s5!=s1
   s6==s1

C. s2==s1
   s3==s1
   s4!=s1
   s5!=s1
   s6!=s1

D. s2!=s1
   s3==s1
   s4!=s1
   s5!=s1
   s6==s1

15. 下面程序段的输出结果是(　　)。

    ```
    StringBuffer buf1=new StringBuffer(20);
    System.out.println(buf1.length()+","+buf1.capacity());
    ```

    A. 0,20　　　　B. 0,null　　　　C. 20,20　　　　D. 0,0

16. 某个应用程序的 main 方法中有下面的语句,输出结果为(　　)。

    ```
    String s="ABCD";
    s.concat("E");
    s.replace('C','F');
    System.out.println(s);
    ```

    A. ABCDEF　　　B. ABFDE　　　C. ABCDE　　　D. ABCD

17. 下面的哪些程序片段可能导致编译错误?(　　)

    A. String s="Gone with the wind";
       String t=" good ";
       String k=s +t;

    B. String s="Gone with the wind";
       String t;
       t=s[3] +"one";

    C. String s="Gone with the wind";
       String standard=s.toUpperCase();

    D. String s="home directory";
       String t=s -"directory";

18. 如果在一个方法中有语句 String[] s＝new String[10];则执行该语句后,下面哪个结论是正确的?(　　)

    A. s[10]为"";
    B. s[9]为 null;
    C. s[0]为未定义
    D. s.length() 为 10

19. 下面的表达式哪个是正确的?(　　)

    A. String s＝"你好";int i＝3; s+＝i;
    B. String s＝"你好";int i＝3; if(i==s){ s+＝i};
    C. String s＝"你好";int i＝3; s=i++;
    D. String s＝null; int i＝(s!＝null)&&(s.length＞0)? s.length():0;

20. 在 Java 语言中,下列表达式返回 false 的是(　　)。

    A. "hello"==="hello"
    B. "hello".equals(new StringBuffer("hello"))
    C. "hello".equals("hello")
    D. "hello".equals(new String("hello"))

21. 在 Java 中,下列语句中正确的是(　　)。

    A. String temp[]＝new String{"a","b","c"};
    B. String temp[]＝{"a" "b" "c"};

C. String temp={"a","b","c"};
D. String temp[]={"a","b","c"};

22. 定义三个整型数组 a1、a2、a3，下面声明正确的语句是（　　）。

  A. intArray[]a1,a2;
   int a3[]={1,2,3,4,5};
  B. int[]a1,a2;
   int a3[]={1,2,3,4,5};
  C. int a1,a2[];
   int a3={1,2,3,4,5};
  D. int[]a1,a2;
   int a3=(1,2,3,4,5);

23. 给出下面代码：

```
public class Person{
  static int arr[]=new int[10];
  public static void main(String a[]){
    System.out.println(arr[1]);
  }
}
```

以下哪个说法是正确的？（　　）

  A. 编译时将产生错误
  B. 编译时正确，运行时将产生错误
  C. 输出 0
  D. 输出空

24. 下面程序段的运行结果是（　　）。

```
String s1="hello";
String s2="hello";
String s3=new String("hello");
System.out.println(s1==s2);
System.out.println(s1==s3);
```

  A. true B. true C. false D. false
   true   false   true   false

25. 对于以下声明：

```
String s1="hello";
String s2="world";
String s3;
```

下面的操作合法的是（　　）。

  A. s3=s1+s2; B. s3=s1-s2;
  C. s3=s1&s2; D. s3=s1&&s2;

26. 下面的程序段执行后输出的结果是（　　）。

```
StringBuffer buf=new StringBuffer("Beijing2008");
buf.replace(0,7,"Hangzhou");
System.out.println(buf.toString());
```

  A. Hangzhou2008 B. Hangzhoug2008
  C. Hangzhou008 D. Beijing2008

27. Java 应用程序的 main 方法中有以下语句,则输出的结果是( )。

    ```
    String  s1="Abc",s2="abc";
    int   i=s1.compareTo(s2);
    System.out.println(i);
    ```

    A. false　　　　　B. -32　　　　　C. 32　　　　　D. 0

5.3.3　程序阅读题

1. 仔细阅读下面的程序代码,回答下面问题。

```
class TestArraySum{
    int a[]=new int[10];
    public TestArraySum(){
        for(int i=0;i<a.length;i++){
            a[i]=2*i+1;
        }
    }
    public int arraySum(int[] a){
        int sum=0;
        for(int i=0;i<a.length;i++){
            sum=a[i]+sum;
        }
        return sum;
    }
    public static void main(String args[]){
        TestArraySum as=new TestArraySum();
        System.out.println(as.arraySum(as.a));
    }
}
```

问:(1) arraySum 方法的功能是什么?

(2) 程序的运行结果是什么?

2. 仔细阅读下面的程序代码,若经编译和运行后,请写出打印结果。

```
enum Season{
    春季,夏季,秋季,冬季
}
public class EnumFor{
  public static void main(String args[ ]){
    for ( Season c : Season.values( ))
      System.out.print(c +", ");
      System.out.println( );
  }
}
```

3. 阅读下面的程序，输出结果是什么？

```java
class TestString{
  public void stringReplace (String text) {
    text=text.replace('j' , 'i');
    text=text+"C";
  }
  public void bufferReplace (StringBuffer text) {
    text.setCharAt(0,'i');
    text=text.append("C");
  }
  public void change(char ch[]){
    ch[0]='Y';
  }
  public static void main(String args[]){
    String str1="java";
    StringBuffer str2=new StringBuffer("java");
    char ch[]={'j','a','v','a'};
    TestString t=new TestString();
    t.change(ch);
    t.stringReplace (str1);
    t.bufferReplace (str2);
    System.out.println(str1);
    System.out.println(str2.toString());
    System.out.println (new String(ch));
  }
}
```

4. 仔细阅读下列程序代码。

```java
class Complex{
    double x,y;
    public Complex(double x,double y) {
        this.x=x;
        this.y=y;
    }
    public static Complex add(Complex a,Complex b) {
        return new Complex(a.x+b.x,a.y+b.y);
    }
}
public class Complexadd {
    public static void main(String args[]) {
      String s=args[0];
      double x,y;
      int p1,p2,len;
      try{
```

```
            p1=s.indexOf("+");    //查找"+"号位置
            if (p1!=-1) {
                String one=s.substring(0,p1);
                String two=s.substring(p1+1);
                p2=one.indexOf(",");
                len=one.length();
                x=Double.parseDouble(one.substring(1,p2));
                y=Double.parseDouble(one.substring(p2+1,len-2));
                Complex c1=new Complex(x,y);
                p2=two.indexOf(",");
                len=two.length();
                x=Double.parseDouble(two.substring(1,p2));
                y=Double.parseDouble(two.substring(p2+1,len-2));
                Complex c2=new Complex(x,y);
                Complex c3=Complex.add(c1,c2);
                System.out.println("结果为("+c3.x+","+c3.y+")");
            }
        }catch(NumberFormatException e) {
            System.out.println("数据格式错!");
        }
    }
}
```

问：(1) 以上代码经编译后,在命令行下输入:

java Complexadd (3,4i)＋(5,8i),则输出结果是什么？

(2) 以上代码经编译后,在命令行下输入:

java Complexadd (3,4i)＋(5,8ai),则输出结果是什么？

5.3.4 编程题

1. 现在有两个双精度浮点型的数组如下：

```
double xs[]={1,9,3,5,6};
double ys[]={10,11,4,7,13};
```

又有一个长方形的数组 Rectangle rects[ ]＝new Rectangle[25]；

要求：用 Rectangle 的构造方法给 rects 数组的每个元素赋值,长和宽分别为 xs 和 ys 的元素；然后按照面积由小到大给 rects 排序。请编写它的程序。

2. 设计一个集合类 Set,然后再编写一个应用程序,创建两个 Set 对象,并利用 Math.random()产生这两个集合中的元素,然后对它们实施"交"、"并"操作。

3. 设计一个矩阵类 Matrix,然后再编写一个 Java 应用程序,创建三个 Matrix 对象 a、b 和 c,初始化 a 和 b,计算矩阵 a 和 b 的乘积,并赋值给矩阵 c,最后输出矩阵 c 的结果。

4. 编写一个 Java 程序 MaxMatrix,找出 4×5 矩阵中值最大的那个元素,显示其值以及其所在的行号和列号。

5. 声明一个二维数组,通过编写程序给其赋值成为一个对角线为 1,其他元素为 0 的 n

阶矩阵,并将其各元素显示出来。

6. 编写一个程序 SortArray.java,对数组 int a[ ]={20,10,50,40,30,70,60,80,90,100}进行从大到小的排序打印输出。

7. 设计一个类 TestArraySum,定义一个含有 10 个元素的 int 类型数组 a,10 个数组元素的值是 11~20,再定义一个方法 arraySum(int[] A),返回数组所有元素的和,最后用 main 方法实现在屏幕上输出数组 a 所有元素的和。

8. 编写一个能对一组字符串(如 hello,world,welcome,hi,hey)按单词表的顺序排序的程序类 WordSort。

9. 编程程序 Tri.java 实现任意输入三条边(a,b,c)后,若能构成三角形且为等腰、等边和直角,则分别输出 DY、DB 和 ZJ,否则输出 YB;若不能构成三角形,则输出 NO。

10. 设有一个英文单词构成的字符串数组,编写程序 StatisticsWord.java 实现。
(1) 统计以字母 w 开头的单词数。
(2) 统计单词中含 or 字符串的单词数。
(3) 统计长度为 3 的单词数。

11. 设有 3 个数组定义如下:

```
String x[]={"zero","one","two", "three","four", "five","six",seven", "eight","nine"};
String y[]={"ten", "eleven","twelve","thirteen","fourteen","fifteen", "sixteen","seventeen","eighteen","nineteen" };
String z[]={"twenty","thirty","fourty","fifty", "sixty","seventy","eighty","ninety"};
```

试编写一个翻译程序 Translate.java,实现用 Java 命令行输入一个小于 100 的整数,将其翻成英文表示,输入英文则输出相应的整数。

例如:

输入 32,输出:thirty two

输入 8,输出:eight

输入 fourteen,输出 14

第 6 章　Java 常用类及接口

6.1　知　识　点

(1) Java API 类库。
(2) Java.lang 包中的常见类。
(3) Java.util 包中的常见类和接口。
重点：Java.lang 包和 Java.util 包中的常见类和接口。
难点：集合的类和接口的应用。

6.2　例　　题

【例 6-1】　欲创建 ArrayList 类的一个实例，此类继承了 List 接口，下列哪个语句是正确的？（　　）
A. ArrayList myList＝new Object（）；
B. List myList＝new ArrayList（）；
C. ArrayList myList＝new List（）；
D. List myList＝new List（）；

【解析】　选项 A 中的 Object 是 Java 中所有类的父类，子类对象可以有父类的类型，反之则不可以。所以，A 选项不能编译。B 选项正确。因为 List 是接口，ArrayList 是实现 List 接口的类，实现接口类的实例对象可以有接口类型声明，反之则不行，C 选项不对。D 选项的 List 是接口，不能实例化。

【例 6-2】　用集合类和接口实现扑克牌的分发。假设有 52 张扑克牌（去掉大小王），实现随机洗牌操作，为参加游戏的人每人生成一手牌，每手牌的牌数是指定的，并将每人分到的牌按花色排序后输出。

【解析】　将去掉大小王的 52 张牌存入 ArrayList 对象，利用集合类 Collections 的 shuffle 方法对 ArrayList 对象中的元素次序打乱（即为洗牌），然后将 ArrayList 对象按序截成四段作为四个玩家的牌，接着利用集合类 Collections 的 sort 方法对在 List 对象中的牌进行排序，最后打印输出四个玩家的牌。

参考代码如下：

```
import java.util.*;
class DealCardDemo {
    public static void main(String args[]) {
        int numHands=4;
        int cardsPerHand=12;
        //生成一副牌(去掉大小王,共 52 张牌)
```

```java
        String[] suit={"♠","♣","♥","♦"};
        String[] rank={"02","03","04","05","06","07","08","09","10","J","Q","K","A"};
        List deck=new ArrayList();
        for (int i=0; i<suit.length; i++)
            for (int j=0; j<rank.length; j++)
                deck.add(suit[i]+rank[j]);
        //随机改变deck中元素的排列次序,即洗牌
        Collections.shuffle(deck);
        /*为四个玩家生成一手牌,并对牌按花色排序后打印输出*/
        for (int i=0; i<numHands; i++){
            List p=dealCard(deck, cardsPerHand);
            Collections.sort(p);
            printCard(p,i+1);
        }
    }
    /*输出某一玩家的牌*/
    public static void printCard(List c,int num){
        System.out.print("玩家"+num+"的牌:");
        for(int i=0;i<c.size();i++){
            String card=(String)c.get(i);
            /*输出时忽略02~09中的0*/
            if (card.charAt(1)=='0'){
                card=card.replace("0","");
            }
            System.out.print(card+" ");
        }
        System.out.println();
    }
    /*为某一玩家分配牌*/
    public static List dealCard(List deck, int n) {
        int deckSize=deck.size();
        List handView=deck.subList(deckSize-n, deckSize);
                                    //从deck中截取一个子链表
        List hand=new ArrayList(handView);
                                    //利用该子链表创建一个链表,作为本方法返回值
        handView.clear();           //将子链表清空
        return hand;
    }
}
```

输出结果为:

玩家1的牌:♠6 ♠7 ♠10 ♠K ♣9 ♣A ♥2 ♥5 ♥7 ♦3 ♦K ♦Q
玩家2的牌:♠3 ♣3 ♣4 ♣7 ♥3 ♥8 ♥2 ♦4 ♦7 ♦8 ♦9 ♦J
玩家3的牌:♠2 ♠4 ♠8 ♠9 ♣2 ♣5 ♣8 ♣J ♥10 ♥J ♥K ♥Q
玩家4的牌:♠5 ♠A ♠Q ♣6 ♣10 ♣K ♣Q ♥4 ♥6 ♥9 ♦10 ♦A

【例 6-3】 用 for 简化写法遍历集合对象元素的实例。

```java
import java.util.*;
class VectorForSimple {
    public static void main(String [] args){
        Vector<String> a=new Vector<String>();
        a.add("Hello");
        a.add("Java");
        a.add("World!");
        for(String str:a){
            System.out.print(str+" ");
        }
    }
}
```

【解析】 本例的代码用 for 简化写法实现遍历集合对象元素。此外，对集合类型的遍历处理还有两种方法。

(1) 通过 size()方法获取集合类型对象的 size，然后通过索引变化用循环可以处理其中的元素。采用这种方法实现本例的代码如下：

```java
import java.util.*;
class VectorForOld{
    public static void main(String [] args){
        Vector<String> a=new Vector<String>();
        a.add("Hello");
        a.add("Java");
        a.add("World!");
        for(int i=0;i<a.size();i++){
            String str=a.get(i);
            System.out.print(str+" ");
        }
    }
}
```

(2) 使用集合类对象的迭代器来遍历其中的每一个元素。调用集合类对象的 iterator()方法可以得到该集合类的迭代器 iterator，然后通过迭代器类 iterator 的 hasNext 和 next 两个方法可以遍历集合对象的元素。采用这种方法实现本例的代码如下：

```java
import java.util.*;
class VectorForIterator {
    public static void main(String [] args){
        Vector<String> a=new Vector<String>();
        a.add("Hello");
        a.add("Java");
        a.add("World!");
        for(Iterator<String> i=a.iterator();i.hasNext();){
            String str=i.next();
```

```
            System.out.print (str+" ");
        }
    }
}
```

6.3 练 习 题

6.3.1 判断题

1. Object 类处于 Java 继承层次中最顶端的类,它是所有 Java 类的父类。 (   )
2. Math 类不能被其他类继承,它的方法和属性均为静态的。 (   )
3. java.util.ArrayList 类采用可变大小的数组实现 java.util.List 接口,并提供了访问数组大小的方法,它的对象会随着元素的增加其容器自动扩大。 (   )
4. java.util.Collections 是集合操作的实用类,提供了集合操作的许多方法,如同步、排序、逆序、搜索等。 (   )
5. java.util.TreeMap 是采用一种有序树的结构实现了 java.util.Map 的子接口 SortedMap,该类按键的升序的次序排列元素。 (   )

6.3.2 选择题

1. 使用以下哪个表达式能产生[20,999]之间的随机整数?(   )
    A. (int)(20+Math.random() * 979)
    B. 20+(int)(Math.random() * 980)
    C. (int)Math.random() * 999
    D. 20+(int)Math.random() * 980
2. 设有如下程序:

```
public class TestInteger {
  public static void main(String args[]){
    Integer intObj=Integer.valueOf(args[args.length-1]);
    int i=intObj.intValue();
    if(args.length>1){
      System.out.println(i);
    }
    if(args.length>0){
      System.out.println(i-1);
    }
    else{
      System.out.println(i-2);
    }
  }
}
```

运行以上程序时,输入命令 java TestInteger 2,则输出结果为( )。
A. —1　　　　　B. 0　　　　　C. 1　　　　　D. 2

3. 下面关于 set 集合处理重复元素的说法正确的是( )。
   A. 如果加入一个重复元素,将抛出异常
   B. 如果加入一个重复元素,add 方法将返回 false
   C. 集合通过调用 equals 方法可以返回包含重复值的元素
   D. 添加重复值将导致编译出错

4. 以下哪个方法是 Vector 类中增加一个新元素的方法?( )
   A. addElement　　B. insert　　　C. append　　　D. addItem

5. 以下哪些方法是 Collection 接口的方法?( )
   A. iterator　　　B. isEmpty　　　C. toArray　　　D. setText

6.3.3　程序阅读题

1. 仔细阅读下面的程序代码,写出运行后的输出结果。

```java
import java.util.Vector;
class TestVector{
    public static void main(String [] args){
        Vector<String>vs=new Vector<String>(16,8);
        vs.add("1");
        vs.add("2");
        vs.add(1,"3");
        System.out.println (vs.get(1));
        if (vs.contains("3"))
            System.out.println ("\"3\"is an element of vs");
        vs.remove("3");
        System.out.println (vs.get(1));
    }
}
```

2. 仔细阅读下面的程序代码,若经编译和运行后,请写出打印结果。

```java
public class Test {
    public static void main(String args[]){
        int [ ] a={1, 2, 3, 4, 5};
        int s=1;
        for (int c: a)
        s * =c;
        System.out.print(s);
    }
}
```

6.3.4　编程题

1. 编写一个程序 ShowDateTime.java,输出当前日期和当前时间。要求当前日期的格式为****年**月**日,如2012年12月20日;当前时间的格式为****-**-** **:**:**,比如2012-12-20 12:12:20。

2. 某班30个学生的学号为20070301~20070330,全部选修了Java程序设计课程,给出所有同学成绩(实现时可用随机数产生,范围为60~100),请编写一个程序StuScore.java将该班的Java程序设计课程成绩按高到低排序打印输出。

要求分别用List和Map来实现,打印的成绩表包括学号、姓名、性别、成绩,如下所示。例如:

```
20120324    张 三    男    95
20120310    李丽芳    女    90
20120302    王小五    男    88
```

3. 编写一个Java成绩统计程序,输入一个班的Java成绩(含姓名和分数),统计出平均分、最高分、最低分,并打印成绩单。由于编程时人数不确定,所以要求使用Vector或ArrayList完成上述程序。要求学生成绩用一个类Mark表示,其成员变量包含考生姓名、成绩。输入的姓名为end时,程序结束。主类的名称叫做GradeStatistic,将Mark设计为GradeStatistic的内部类。

提示:键盘输入采用Scanner类,实例化Scanner采用下述语句:

```
Scanner sc=new Scanner(System.in);
```

然后通过sc调用Scanner的各种方法输入相应的数据。如用next()方法读入一个字符串,nextInt()读入一个整数,nextDouble()方法读入一个双精度数。输入完毕后用sc.close()释放sc对象。Scanner详细用法请参阅JDK的API帮助文档。

第7章 异常处理

7.1 知 识 点

(1) Java 中的异常类。
(2) 异常处理模式 try-catch-finally。
(3) 重新抛出异常。
(4) 自定义异常类。
重点：异常处理模式 try-catch-finally，自定义异常类。
难点：异常的处理机制，自定义异常类。

7.2 例 题

【例 7-1】 异常处理的实例。

```java
public class ExceptionExample{
    public static void main(String[] args) {
        try{
            int b=12;
            int c;
            for (int i=2;i>=-2;i--){
                c=b/i;
                System.out.println("i="+i);
            }
        }catch(ArithmeticException arithmeticException) {
            System.out.println("捕获了一个零除异常");
        }catch(Exception arithmeticException) {
            System.out.println("捕获了一个未知类型的异常");
        }finally{
            System.out.println("异常处理结束");
        }
    }
}
```

【解析】 异常(Exception)是正常程序流程所不能处理或没有处理的异常情况或异常事件，比如算术运算被 0 除、数组下标越界等。Java 采用 try-catch-finally 语句捕获并处理异常，使用的格式如下：

```
try{
    …//被监视的代码块
```

```
        }
        catch(异常类1  对象名1){
            …//异常类1的异常处理代码块
        }
        catch(异常类2  对象名2){
            …//异常类2的异常处理代码块
        }
        finally{
            …//在try块结束前被执行的代码块
        }
```

其中,try语句后必须跟一个catch或finally语句块,发生异常时try语句块内的后续语句不再执行;try中的语句一旦出现异常,catch按照次序进行匹配检查处理,找到一个匹配者,不再找其他。因此,catch的排列要按照捕获异常对象的子类异常排在前、将父类异常排在后的原则。finally是异常处理的统一出口,常用来实现资源释放,如文件关闭、关闭数据库连接等。除遇到System.exit()强制退出程序外,finally语句块无论是否发生异常都要执行。

因此,本例程序执行后输出结果如下:

i=2
i=1
捕获了一个零除异常
异常处理结束

【例7-2】 下面的程序经编译后,运行结果是什么?(   )

```
1.   class Person{
2.       String name;
3.       char sex;
4.       int age;
5.       public String toString(){
6.           return "name:"+name+" sex:"+sex+" age:"+age;
7.       }
8.   }
9.   class Student extends Person{
10.      String major;
11.      public String toString(){
12.          return "name:"+name+" sex:"+sex+" age:"+age+" major:"+major;
13.      }
14.      public static void main(String a[]){
15.          Student stu[ ]=new Student[20];
16.          for(Student s:stu){
17.              s.name="无名氏";
18.              s.sex='男';
19.              s.age=20;
```

```
20.            s.major="软件工程";
21.            System.out.println(s.toString());
22.        }
23.    }
24. }
```

A. 11 行编译出错

B. 14 行编译出错

C. 编译正确,运行时出现异常

D. 编译正确,运行结果为打印 20 个学生信息

【解析】 第 11 行的 toString 方法覆盖了 Person 类中的 toString 方法,编译不会出错;第 14 行为 main 方法,作为程序执行的入口,main 方法的声明一般是 public static void main(String args[]),这里唯一能修改的是参数变量名 args,所以 14 行的编译也不会出错;纵观整个程序代码,编译都不会出错,但执行时第 17 行会抛出异常 java.lang.NullPointerException,原因是第 15 行仅创建了 Student 的一个数组 stu,但并没有对数组中的成员实例化,导致第 17 行对象 s 为 null。因此,本例选 C。

7.3 练 习 题

7.3.1 选择题

1. 下列哪种操作不会抛出异常?(　　)

　　A. 打开不存在的文件　　　　　　B. 用负数索引访问数组

　　C. 浮点数除以 0　　　　　　　　D. 浮点数乘 0

2. 如果一个程序中有多个 catch 语句,则程序会按如下哪种情况执行?(　　)

　　A. 找到合适的异常类型后继续执行后面的 catch 语句

　　B. 找到每个符合条件的 catch 都执行一次

　　C. 找到合适的异常类型处理后就不再执行后面的 catch 语句

　　D. 对每个 catch 都执行一次

3. 对于异常处理语句 try-catch-finally,下面哪个说法是正确的?(　　)

　　A. 如果有多个 catch 语句,对所有的 catch 都会执行一次

　　B. 如果有多个 catch 语句,对每个符合条件的 catch 都会执行一次

　　C. 多个 catch 的情况下,异常类的排列顺序应该是父类在前,子类在后

　　D. 一般情况下,finally 部分都会被执行一次

4. 程序员将可能发生异常的代码放在(　　)块中,无论如何都需要执行的代码在(　　)块。

　　A. catch、try　　　　　　　　　　B. try、finally

　　C. try、exception　　　　　　　　D. try、final

5. Java 程序运行时会自动检查数组的下标是否越界,如果越界,会抛出下面哪一个异常?(　　)

　　A. SQLException　　　　　　　　B. IOException

C. ArrayIndexOutOfBoundsException　　D. SecurityManager

6. 关于异常处理的语法 try-catch-finally，下列描述正确的是（　　）
   A. try-catch 必须配对使用
   B. try 可以单独使用
   C. try-finally 必须配对使用
   D. 在 try-catch 后如果定义了 finally，则 finally 一般都会执行

7. 给定下面的代码片段：
```java
public class Animal{
    public void cat(){
    try{
        method();
    }
    catch(ArrayIndexOutBoundsException e){
        System.out.println("Exception1");
    }
    catch(Exception e){
        System.out.println("Exception2");
    }
    finally{
        System.out.println("Hello World!!");
    }
    }
    public void method(){
        //...
    }
    public static void main(String[] args) {
        Animal animal=new Animal();
        animal.cat();
    }
}
```
   如果方法 method 正常运行并返回，程序的输出结果是（　　）。
   A. Hello World　　　　　　　　B. Exception1
   C. Exception2　　　　　　　　D. Hello World!!

8. 如果执行下面的 example 方法时，unsafe()有异常，则输出结果是（　　）。
```java
public void example(){
    try{
        unsafe();
        System.out.print("1");
    }
    catch(Exception e){
        System.out.print("2");
    }
```

```
            finally{
                System.out.println("3");
            }
            System.out.print("4");
    }
```

  A. 123   B. 234   C. 23   D. 34

9. 考虑下列 Java 代码，编译和运行后的情况是（   ）。

```
1.   public class ArrayIndexOutOfBoundsExceptionDemo{
2.     public static void main(String args[ ]){
3.       int i=0;
4.       String greetings[ ]={"Hello World!","Hello!","HELLO WORLD!!"};
5.       while (i<4){
6.         System.out.println(greetings[i]);
7.         i++;
8.       }
9.     }
10.  }
```

  A. 第 4 行出现编译错误

  B. 第 6 行出现编译错误

  C. 编译正确，但运行程序时第 6 行出现异常

  D. 编译正确，运行程序能正常输出信息

10. 下面的程序编译后用以下命令运行：java ExceptionDemo hello，运行的情况是（   ）。

```
public class ExceptionDemo{
    public static void main(String args[ ]){
        for(int i=0;i<2;i++){
            System.out.println(args[i]);
        }
    }
}
```

  A. 先输出 hello，发生异常，异常类型是 ArithmeticException

  B. 发生异常，异常类型是 IOException

  C. 先输出 hello，然后发生异常，异常类型是 ArrayIndexOutOfBoundsException

  D. 程序正常输出信息 hello

7.3.2 程序阅读题

1. 写出下面程序标记为 1～6 的执行顺序。

```
public class UsingExceptions {
  public static void main(String args[] ){
    try{
```

```
        method1();                                          //1
    }
    catch(Exception e){
      System.err.println(e.getMessage());                   //2
    }
    finally{
      System.out.println("Program is end!");                //3
    }
  }
  public static void method1() throws Exception{
    method2();                                              //4
  }
  public static void method2() throws Exception{
    method3();                                              //5
  }
  public static void method3() throws Exception{
    throw new Exception( "Exception thrown in method3" );   //6
  }
}
```

2. 仔细阅读下面的程序代码。

```
1.   class TestException{
2.     public static void main(String []args){
3.       try{
4.         method();
5.       }
6.       catch(Exception e){
7.         System.out.print('m');
8.       }
9.       System.out.println('n');
10.    }
11.    static void createException(){
12.      throw new ArrayIndexOutOfBoundsException();
13.    }
14.    static void method(){
15.      try{
16.        createException();
17.        System.out.print('a');
18.      }
19.      catch(ArithmeticException e){
20.        System.out.print('b');
21.      }
22.      finally{
23.        System.out.print('c');
24.      }
```

```
25.         System.out.print('d');
26.     }
27. }
```

（1）以上程序代码若经编译和运行后,输出结果是什么？

（2）若第 19 行的 ArithmeticException 改为 ArrayIndexOutOfBoundsException,则输出结果又是什么？

3. 阅读下面的程序 TestMonth.java：

```
public class TestMonth{
    public static void main(String []args){
        try{
            int month=Integer.parseInt(args[0]);
            if(month<0||month>12){
                throw new ArithmeticException("没有"+month+"月份!");
            }
            System.out.println("您输入的月份为"+month+"月份");
        }
        catch(ArrayIndexOutOfBoundsException e){
            System.out.println("请输入月份!");
        }
        catch(ArithmeticException e){
            System.out.println("捕获 ArithmeticException 异常");
            System.out.println(e.getMessage());
        }
    }
}
```

已知 ArrayIndexOutOfBoundsException 和 ArithmeticException 都是 java.lang.* 下的异常类,编译 TestMonth.java 后,用 java TestMonth 13 的运行结果是什么？

4. 仔细阅读下面的程序代码,若经编译和运行后,请写出打印结果。

```
class NoTriangleException extends Exception{
    NoTriangleException(){}
    NoTriangleException(String str){
        super(str);
    }
}
class Triangle{
    int x, y, z;                              //定义三角形的三条边
    public Triangle(){
        x=y=z=1;                              //初始化三角形三条边的长度为 1
    }
    public Triangle(int x, int y, int z){
        this.x=x;
        this.y=y;
```

```java
            this.z=z;
        }
        public double getArea(){
            double d= (x +y +z) / 2.0;
            try{
                if(x<0||y<0||z<0||x+y<=z||x+z<=y||y+z<=x )
                    throw new NoTriangleException("不能构成三角形");
            }catch(NoTriangleException e){
                System.out.println("边长分别为"+x+","+y+","+z+","+e.getMessage());
                return 0;
            }
            return Math.sqrt(d * (d-x) * (d-y) * (d-z));
        }
    }
    public class TestTriangle {
        public static void main(String args[]){
            Triangle t=new Triangle(5, 5, 10);
            double area=t.getArea();
            if(area>0.0){
                System.out.println("面积为"+area);
            }
        }
    }
```

5. 仔细阅读下面的程序代码，若经编译和运行后，请写出打印结果。

```java
class myException extends Exception{}
public class Sample{
    public void foo(){
        try{
            System.out.print(a);
            bar();
            System.out.print(b);
        }
        catch(myException e){
            System.out.print(c);
        }
        finally{
            System.out.print(d);
        }
    }
    public void bar() throws myException{
        throw new myExcepticn();
    }
    public static void main(String args[]){
        Sample s=new Sample();
```

```
        s.foo();
    }
}
```

6. 写出下列程序代码的运行结果：

```java
class MinusException extends Exception{
    int number;
    public MinusException (String msg, int i){
        super(msg);
        this.number=i;
    }
}
class Salary{
  private String name;
  private int salary;
  public Salary(String n, int s ) throws MinusException{
    this.name=n;
    if (s<0) throw new MinusException("工资不能为负!",s);
    this.salary=s;
  }
  public void print(){
    System.out.println(name+"的工资为"+salary+"元。");
  }
}
class TestSalary{
    public static void main(String [] args){
        try {
            Salary s1=new Salary("张三",1000);
            s1.print();
            Salary s2=new Salary("李四",-10);
            s2.print();
        }
        catch(MinusException e){
            System.out.println("异常:"+e.getMessage());
        }
    }
}
```

7. 写出下列程序代码的运行结果。

```java
public class TestFactor{
    int n;
    public void printInfo(){
        try{
            System.out.println("求 12 的因子个数");
            n=f(12);
```

```
            System.out.println("12 的因子个数为"+n);
        }catch(Exception e){
            System.out.println("求 12 的因子个数有误");
        }
        finally{
            System.out.println("结束");
        }
    }
    public int f(int n){
        int y=0;
        for(int i=n;i>=0;i--){
            if(n%i==0)y++;
        }
        return y;
    }
    public static void main(String args[]){
        TestFactor tf=new TestFactor();
        tf.printInfo();
    }
}
```

7.3.3 编程题

1. 编写一个程序,从命令行参数输入 10 个数作为学生成绩,需对成绩进行有效性判断,若成绩有误则通过异常处理显示错误信息,并将成绩按高到低排序打印输出。提示:如果输入数据不为整数,要捕获 Integer.parseInt()产生的异常,显示"请输入成绩",捕获输入参数不足 10 个的异常,显示"请输入至少 10 个成绩"。

2. 编写一个银行类 Bank,要求如下:
(1) 变量 balance 为存款余额。
(2) deposite()方法为存款操作。
(3) withdrawa()方法为取款操作。
(4) getbalawal()方法为获取余额操作,如果银行余额少于取款额时,显示当前余额,并告知不能取款的提示,否则显示取款成功的信息。要求用自定义异常来处理余额不足的问题。

3. 自定义两个异常,一个叫 EmptyStackException,其默认的异常信息是"堆栈空!";另一个叫做 FullStackException,其默认的异常信息是"堆栈满!"。

4. 应用泛型编写一个堆栈类 MyStack,元素可以是任何引用类型的对象,MyStack 的构造方法指定堆栈的大小,MyStack 类中包括 push()、pop()、isEmpty()、isFull()方法。push()方法申明抛出 FullStackException 异常,pop()方法申明抛出 EmptyStackException。异常编写一个测试类 TestMyStack,创建一个元素为 String,大小为 5 的堆栈,并使用其方法进行进栈出栈等操作。

第8章 流和文件

8.1 知 识 点

(1) 流的基本概念。
(2) 字节流,包括 InputStream 和 OutputStream,FileInputStream 和 FileOutputStream,DataInputStream 和 DataOutputStream。
(3) 字符流,包括 Reader 和 Writer,FileReader 和 FileWriter,BufferedReader 和 BufferedWriter。
(4) 文件,包括文件类 File 和随机访问文件类 RandomAccessFile 以及文件过滤接口 FileFilter 和 FilenameFilter。
(5) 对象序列化。

重点:字节流和字符流相关类,文件类 File 和随机访问文件类 RandomAccessFile,对象序列化。

难点:读写文件的应用,对象序列化。

8.2 例 题

【例 8-1】 编写一个复制任意类型文件的类程序 CopyFile.java,该类就只有一个方法 copy(),方法声明如下:

public boolean copy(String fromFileName, String toFileName,boolean override){…}

其中,参数 1:fromFileName 是源文件名;参数 2:toFileName 是目标文件名;参数 3:override 是目标文件存在时是否覆盖。若文件复制成功,则 copy()方法返回 true,否则返回 false。

【解析】 复制文件的过程是先读取被复制的文件内容,然后将内容写入另一个文件,可采用 FileInputStream 和 FileOutputStream 分别读写文件。本例中的 copy 方法需要判断目标文件是否覆盖,如果参数 override 值为 true,覆盖目标文件,否则就不覆盖。本例的参考代码如下:

```
1.    class CopyFile{
2.        public boolean copy (String fromFileName, String toFileName, boolean override){
3.            File fromFile=new File(fromFileName);      //创建源文件的 File 对象
4.            File toFile=new File(toFileName);           //创建目标文件的 File 对象
5.            /*判断源文件是否存在、是否为文件、是否可读*/
6.            if(!fromFile.exists()||!fromFile.isFile()
                ||!fromFile.canRead()){
```

```
7.          return false;
        }
8.      if (toFile.isDirectory()) {              //判断目标文件是否为目录
9.        toFile=new File(toFile, fromFile.getName());
10.     }
11.     if (toFile.exists()) {                   //判断目标文件是否存在
12.       /*判断目标文件是否可写以及是否可覆盖*/
13.       if (!toFile.canWrite()||override==false) {
14.          return false;
15.       }
16.     }
17.     else{
18.       /*判断目标文件所在的目录是否存在以及是否可写*/
19.       String parent=toFile.getParent();
20.       if (parent==null) {
21.         parent=System.getProperty("user.dir");
22.       }
23.       File dir=new File(parent);
24.       if (!dir.exists()||dir.isFile()||!dir.canWrite()) {
25.          return false;
26.       }
27.     }
28.     FileInputStream from=null;
29.     FileOutputStream to=null;
30.     try{
31.       from=new FileInputStream(fromFile);
32.       to=new FileOutputStream(toFile);
33.       byte[] buffer=new byte[4096];
34.       int bytes_read;
35.       /*从源文件读取的内容写入目标文件中*/
36.       while ( (bytes_read=from.read(buffer)) !=-1) {
37.          to.write(buffer, 0, bytes_read);
38.       }
39.       return true;
40.     }
41.     catch (IOException e) {
42.       return false;
43.     }
44.     finally {
45.       if (from !=null) {
46.         try {
47.           from.close();
48.         }
49.         catch (IOException e) { }
50.       }
```

```
51.         if (to !=null) {
52.           try {
53.             to.close();
54.           }
55.           catch (IOException e) { }
56.         }
57.       }
58.   }
59. }
```

【例 8-2】 从键盘输入文字存入文件,再读出加上行号后打印在屏幕上。

【解析】 读取键盘的输入用标准输入流类 System.in,该类是字节输入流。若键盘的输入内容有中文字符并需逐行读取,可采用带缓冲的字符输入流 BufferedReader 来读取。本例的参考代码如下:

```
1.  import java.io.*;
2.  public class BufferDemo{
3.    public static void main(String []args){
4.      String f="f.txt";
5.      String str="";
6.      int i=0;
7.      try{
8.        /*用 BufferedReader 封装标准输入流类 System.in*/
9.        BufferedReader keyIn=new BufferedReader(new InputStreamReader(System.in));
10.       BufferedWriter bw=new BufferedWriter(new FileWriter(f));
11.       BufferedReader br=new BufferedReader(new FileReader(f));
12.       System.out.println("Please input file text:");
13.       /*如果从键盘输入的内容不是"exit",则按行写入文件中*/
14.       while(!(str=keyIn.readLine()).equals("exit")){
15.         bw.write(str,0,str.length());
16.         bw.newLine();
17.       }
18.       bw.close();
19.       /*按行读出文件内容并加上行号后打印在屏幕上*/
20.       while((str=br.readLine())!=null){
21.         i++;
22.         System.out.println(i+": "+str);
23.       }
24.     }catch(IOException e){ }
25.   }
26. }
```

使用输入输出流类的注意事项如下。

(1) 复制文件,可用字节输入输出流 FileInputStream 和 FileOutputStream 实现,见例 8-1。

(2) 读写文件内容,可用 BufferedReader 和 BufferedWriter 或 PrintWriter 实现,见

例8-2。

(3) 先用输出流写文件,然后用输入流读该文件时,使用的输入输出流应成对使用,比如 FileInputStream 和 FileOutputStream,DataInputStream 和 DataOutputStream,BufferedReader 和 BufferedWriter 或 PrintWriter。不能写文件的输出流用 FileOutputStream,而读文件的输入流采用 BufferedReader,这样会导致读取的数据乱码。

(4) 可用 BufferedReader 读取键盘输入内容,语句如下:

```
BufferedReader keyIn=new BufferedReader(new InputStreamReader(System.in));
keyIn.readLine();
```

(5) 可用类 java.util.Scanner 读取键盘输入内容,使用方法详见 JDK API。

【例8-3】 编写一个程序,统计给定文件中包含的每个单词出现的频率,并按单词表的顺序显示统计结果。

【解析】 本例的设计思路是读取给定文件所有内容存入字符串,然后,分离出单词,并对每个单词创建一个对象(包含该单词和出现的次数),接着,将这些单词对象逐个放入 ArrayList 对象中,并排序。放入 ArrayList 对象过程中,判断该单词是否已存在,若存在,则出现的次数加1;最后,将单词和出现的次数打印输出。本例的参考代码如下:

```
1.   import java.util.*;
2.   import java.io.*;
3.   public class TestWordCount{
4.     private String word;
5.     private int wordNum;                         //单词出现的次数
6.     public TestWordCount(String wordStr,int num){
7.       word=wordStr;
8.       wordNum=num;
9.     }
10.    public static void main(String[] args){
11.      String fileName="WordCount.txt";
12.      /*读取文件 WordCount.txt 中的内容*/
13.      try{
14.      BufferedReader br=new BufferedReader(new FileReader(fileName));
15.      String line=br.readLine();
16.      StringBuffer fileContent=new StringBuffer();
17.      while(line!=null){
18.        fileContent.append(line);
19.        line=br.readLine();
20.      }
21.      wordCount(fileContent.toString());
22.      br.close();
23.      }
24.      catch(Exception ex){
25.        ex.printStackTrace();
26.      }
27.    }
```

```
28.     /*分离出所有单词,并统计单词的次数*/
29.     static void wordCount(String fileContent){
30.         String str=fileContent;
31.         String str1=str.replaceAll("[^a-zA-Z\']"," ");
32.         StringTokenizer st=new StringTokenizer(str1," ");          //分离出单词
33.         int j=st.countTokens();
34.         ArrayList wordcount=new ArrayList();
35.         for(int i=0;i<j;i++){
36.             wordcount=wordSort(wordcount,st.nextToken().toLowerCase());
37.         }
38.         //输出 wordCount 结果
39.         for(int i=0;i<wordcount.size();i++){
40.             System.out.println(((TestWordCount)wordcount.get(i)).word +": " +
                    ((TestWordCount)wordcount.get(i)).wordNum);
41.         }
42.     }
43.     /*将单词按序统计次数*/
44.     static ArrayList wordSort(ArrayList a,String aWord){
45.         TestWordCount wordArray=new TestWordCount(aWord,1);
46.         if(a.size()<1) a.add(wordArray);//
47.         else{
48.             for(int i=a.size()-1;i>=0;i--){
49.                 int flag=aWord.compareTo(((TestWordCount)a.get(i)).word);
50.                 if(flag==0){                                  //说明该单词已统计
51.                     wordArray.wordNum=((TestWordCount)a.get(i)).wordNum+1;
52.                     wordArray.word=aWord;
53.                     a.set(i,wordArray);
54.                     break;
55.                 }//if(flag==0)
56.                 if(flag>0){
57.                     a.add(i+1,wordArray);
58.                     break;
59.                 }//if(flag>0)
60.                 if(flag<0&&i==0)
61.                     a.add(0,wordArray);
62.             }//for
63.         }//else
64.         return a;
65.     }
66. }
```

【例 8-4】 随机访问文件的实例,要求在文件中写入浮点类型的 $1,2,3,\cdots,10$ 共 10 个数,然后将第三个数 3.0 修改为 0.0,最后读出文件中的 10 个数并打印。

【解析】 类 java.io.RandomAccessFile 可以实现对文件随机访问,它既可以读取文件任意位置的内容,也可以在任意位置写入数据。它的一个构造方法是 RandomAccessFile

(String name,String mode)，其中 name 是文件名，mode 是对文件的访问模式：r 表示读，w 表示写，rw 表示既可读又可写。本例的参考代码如下：

```java
1.  import java.io.IOException;
2.  import java.io.RandomAccessFile;
3.  public class RandomAccessFileDemo{
4.    public static void main(String args[ ]){
5.      try{
6.        /*创建一个可读可写的 RandomAccessFile 对象 f */
7.        RandomAccessFile f=new RandomAccessFile("test.txt", "rw");
8.        int   i;
9.        double  d;
10.       /*向文件中写入浮点类型的 1,2,3,…,10 共 10 个数 */
11.       for (i=1; i<=10; i++){
12.         f.writeDouble(i);
13.       }
14.       f.seek(16);                //将文件指针转到第三个数据的位置
15.       f.writeDouble(0);          //将当前位置的数据修改为 0.0
16.       f.seek(0);                 //重新将文件指针指向文件初始位置
17.       /* 读出文件中的 10 个数并打印 */
18.       for (i=0; i<10; i++){
19.         d=f.readDouble();
20.         System.out.println("[" +i +"]: " +d);
21.       }
22.       f.close();
23.     }
24.     catch (IOException e){
25.       System.err.println("发生异常:" +e);
26.       e.printStackTrace();
27.     }
28.   }
29. }
```

【例 8-5】 编写程序实现学生数据的序列化存于文件中。

【解析】 序列化的过程就是对象写入字节流和从字节流中读取对象。将对象状态转换成字节流之后，可以用 java.io 包中的各种字节流类将其保存到文件中，或通过管道传到另一线程中，或通过网络连接将对象数据发送到另一主机。对象序列化在 RMI、Socket、JMS、EJB 中都有应用。为实现对象数据的序列化读取，首先实现 java.io.Serializable 接口的类，然后采用 java.io.ObjectOutputStream 将对象写入字节流，最后用 java.io.ObjectInputStream 从字节流重构对象。本例的参考代码如下：

(1) 首先实现 java.io.Serializable 接口的类 Student：

```java
1.  import java.io.Serializable;
2.  public class Student implements Serializable{
3.    static final long serialVersionUID=123456L;
```

```
4.      String name;
5.      int id;
6.      int height;
7.      public Student( String name, int id, int height ){
8.          this.name=name;
9.          this.id=id;
10.         this.height=height;
11.     }
12.     public void output( ){
13.         System.out.println("姓名：" +name);
14.         System.out.println("学号：" +id);
15.         System.out.println("身高：" +height);
16.     }
17. }
```

（2）采用 java.io.ObjectOutputStream 将 Student 对象数据写入 object.dat。

```
1.  import java.io.FileOutputStream;
2.  import java.io.ObjectOutputStream;
3.  public class WriteObject{
4.      public static void main(String args[ ]){
5.          try{
6.              ObjectOutputStream f=new ObjectOutputStream(
7.                  new FileOutputStream("object.dat"));
8.              Student s=new Student( "张三", 201213001, 172);
9.              f.writeObject(s);            //将 Student 对象 s 写入文件
10.             s.output( );
11.             f.close( );
12.         }
13.         catch (Exception e){
14.             System.err.println("发生异常:" +e);
15.             e.printStackTrace( );
16.         }
17.     }
18. }
```

（3）用 java.io.ObjectInputStream 从 object.dat 读出 Student 对象数据。

```
1.  import java.io.FileInputStream;
2.  import java.io.ObjectInputStream;
3.  public class ReadObject{
4.      public static void main(String args[ ]){
5.          try{
6.              ObjectInputStream f=new ObjectInputStream(
7.                  new FileInputStream("object.dat"));
8.              /*从文件中读出序列化对象,并重构成学生对象*/
9.              Student s= (Student)(f.readObject( ));
```

```
10.        s.output();
11.        f.close();
12.     }
13.     catch (Exception e){
14.       System.err.println("发生异常:" +e);
15.       e.printStackTrace();
16.     }
17.   }
18. }
```

8.3 练 习 题

8.3.1 判断题

1. 利用 File 对象可以判断一个文件或目录是否存在。 ( )
2. System.in 是标准输入流,能用 read 方法读取键盘的输入。 ( )
3. 数据流就是数据通信通道,指在计算机的输入输出之间的数据序列。 ( )
4. 数据流就是从源到目的字节的有序序列,包括输入流和输出流。 ( )
5. 在 Java 中,类 Java.io.File 虽然不能直接处理文件内容,但可以通过类 Java.io.File 修改文件名。 ( )

8.3.2 选择题

1. 以下哪一项不是 File 类的功能?( )
   A. 创建文件　　　B. 创建目录　　　C. 删除文件　　　D. 复制文件
2. 下面哪个不是 InputStream 类中的方法?( )
   A. int read(byte[])   B. void flush()
   C. void close()   D. int available()
3. 创建 BufferedInputStream 对象时,下面哪个类的对象可作为参数?( )
   A. File   B. BufferedOutputStream
   C. FileInputStream   D. FileOutputStream
4. 要从文件 file.dat 中读出第 10 个字节到变量 c 中,下列哪个语句段适合?( )
   A. FileInputStream in=new FileInputStream("file.dat");
      in.skip(9);
      int c=in.read();
   B. FileInputStream in=new FileInputStream("file.dat");
      in.skip(10);
      int c=in.read();
   C. FileInputStream in=new FileInputStream("file.dat");
      int c=in.read();

D. RandomAccessFile in=new RandomAccessFile("file.dat");
   in.skip(9);
   int c=in.readByte();

5. 关于 Java 输入输出流,下列说法正确的是(　　)。
   A. FileReader 类和 FileInputStream 类都是按照字节读取的
   B. FileReader 类和 FileInputStream 类都是按照字符读取的
   C. FileReader 类是按字节读取的,FileInputStream 类是按字符读取的
   D. FileReader 类是按字符读取的,FileInputStream 类是按字节读取的

6. 下面哪一个流属于字符流?(　　)
   A. InputStream   B. FileInputStream
   C. DataInputStream   D. FileReader

7. 以下哪个是创建 RandomAccessFile 对象的正确方法?(　　)
   A. new RandomAccessFile("data", "r");
   B. new RandomAccessFile("r", "data");
   C. new RandomAccessFile("data", "read");
   D. new RandomAccessFile("read", "data");

8. 下列哪一个对象引用 f 既能从文件 file.txt 读取数据,又能向文件 file.txt 写入数据?(　　)
   A. File f=new File("file.txt");
   B. RandomAccessFile f=new RandomAccessFile("file.txt","rw");
   C. FileOutputStream f=new FileOutputStream("file.txt");
   D. FileInputStream f=new FileInputStream("file.txt");

9. File 类的构造方法 public File(String parent,String child)中,参数 child 是(　　)。
   A. 子文件夹名   B. 子文件夹对象名
   C. 文件名   D. 文件对象名

10. 编译和运行下面的程序,并从键盘输入 12345,则输出的结果是(　　)。

```
import java.io.*;
public class Test{
  public static void main(String args[]) throws IOException{
    BufferedReader buf=new BufferedReader(
            new InputStreamReader(System.in));
    String  str=buf.readLine();
    int x=Integer.parseInt(str);
    System.out.println(x%1000);
  }
}
```

   A. 45   B. 345   C. 123   D. 12345

11. 下面程序段的功能是(　　)。

```
File file=new File("E:\\xxx\\yyy");
```

```
        file.mkdirs();
```
  A. 在当前目录下生成子目录\xxx\yyy B. 生成目录 E:\xxx\yyy
  C. 在当前目录下生成文件 xxx.yyy D. 以上说法都不对

12. 用"new FileOutputStream("data.txt",true)"创建一个 FileOutputStream 的实例对象,则下面哪个说法是错误的?( )

  A. 如果文件 data.txt 不存在,也不一定会抛出 IOException 异常
  B. 如果文件 data.txt 不存在,则可能会新建文件 data.txt
  C. 如果文件 data.txt 存在,则将覆盖掉文件中原有的内容
  D. 如果文件 data.txt 存在,则从文件的末尾开始添加新内容

8.3.3 程序阅读题

1. 阅读下面的程序,写出带划线语句或注释,并写出该程序的功能。

```
1.   import java.io.*;
2.   public class Test {
3.     public static void main(String args[]) {
4.       scanFiles("c:/");
5.     }
6.     public static void scanFiles(String path) {
7.       if (path==null)
8.         return;
9.       File f=new File(path);        //_____
10.      if (!f.exists())
11.        return;
12.      if (f.isFile())               //_____
13.        System.out.println(f.getAbsolutePath());
14.      else{
15.        File dir[]=f.listFiles();
16.        for (int i=0; i<dir.length; i++)
17.          scanFiles(dir[i].getAbsolutePath());
18.      }
19.    }
20.  }
```

2. 阅读下面的程序 Test.java,先填写空格的内容,然后写出运行结果。

```
1.   import java.io.*;
2.   public class Test{
3.     public static void main(String argv[]){
4.       _____;    //创建 Test 对象,对象名为 t
5.       System.out.println(t.fliton());
6.     }
7.     public int fliton(){
8.       try{
```

```
                //第 9 行的含义是：_____
9.          FileInputStream din=new FileInputStream("test.txt");
10.         din.read();
11.     }catch(IOException  ioe){   //第 11 行的含义是：_____
12.         System.out.println("one");
13.         return -1;
14.     }
15.     finally{
16.         System.out.println("two");
17.     }
18.     return 0;
19.   }
20. }
```

如果文件 test.txt 与 Test.java 在同一个目录下，test.txt 中仅有一行字符串"hello world!"，运行结果是什么？

3. 下面的程序 ReadFileContent.java 实现的功能是读取当前目录下的 Test.txt 文件内容（内容含有中文字），将该文件的内容按行读取出来，并在每行前面加上行号后写入当前目录的 myTest.txt 文件中。请将下列(1)~(5)的语句补充完整。

```
1.  import     (1)               ;
2.  class ReadFileContent{
3.    public static void main(String args[]){
4.      int i=0;
5.      try{
6.        BufferedReader br=new BufferedReader(    (2)    );
7.        String line=     (3)    ;
8.        PrintWriter out=new PrintWriter(     (4)     );
9.        while(line!=null){
10.         i++;
11.         out.println(i+" "+line);
12.         line=br.readLine();
13.       }
14.       br.close();
15.       out.close();
16.     }
17.       (5)     {
18.       e.printStackTrace();
19.     }
20.   }
21. }
```

4. 当前目录不存在名为 Hello.txt 的文件，执行下面代码的输出结果是什么？

```
1.  import java.io.*;
2.  public class ReadFile{
```

```
3.    public static void main(String argv[]){
4.      ReadFile r=new ReadFile();
5.      System.out.println(r.amethod());
6.    }
7.    public int amethod(){
8.      try{
9.        FileInputStream file=new FileInputStream("Hello.txt");
10.     }
11.     catch(FileNotFoundException e){
12.       System.out.println("文件没找到");
13.       return -1;
14.     }
15.     catch(IOException e){
16.       System.out.println("出现 IO 异常");
17.     }
18.     finally{
19.       System.out.println("退出");
20.     }
21.     return 0;
22.   }
23. }
```

5. 下面的程序是统计单词 hello 在一篇英文文章(保存在文件 C:/article.txt)中出现的次数，统计时忽略单词的大小写，统计结果在屏幕上打印出来的格式为：单词＊＊＊在文章＊＊＊中出现的次数为：＊＊＊。请将划线上(1)~(4)的语句补充完整。

```
1.  import java.io.BufferedReader;
2.  import java.io.FileReader;
3.  import java.io.IOException;
4.  class WordCount{
5.    int getWordCont(String articleName,String word){
6.      int count=0;
7.      String sentence;
8.      try{
9.        BufferedReader br=new BufferedReader(____(1)____);
10.       word=word.toLowerCase();
11.       while((sentence=____(2)____)!=null){
12.         sentence=sentence.toLowerCase();
13.         int index=sentence.indexOf(word);
14.         while(index>=0){
15.           count++;
16.           //若该行中还含有该单词,则继续查找
17.           index=____(3)____;
18.         }
19.       }
20.     }catch(IOException e){
```

```
21.            System.out.println("error!");
22.        }
23.        return count;
24.    }
25.    public static void main(String args[]){
26.        String article="c:/article.txt";
27.        String word="hello";
28.        _____(4)_____;
29.        System.out.println("单词"+word+"在文章"+article+"中出现的次数为:"+
           w.getWordCount(article,word));
30.    }
31. }
```

8.3.4 编程题

1. 编写一个程序 ReadText.java，按行顺序地读取一个文件 myText.txt 的内容并打印输出。

2. 编写一个检测文件程序 CheckFile.java，判断当前目录下 myText.txt 是否存在，若存在则输出其长度。

3. 编写一个程序 WriteLog.java 实现如下功能：从键盘输入若干行文字（可能包含中文），当最后一行输入 quit♯ 时，退出程序且将输入内容除 quit♯ 外全部存入文件 D:\log.txt 中。

4. 编写一个程序 WordStatistic.java，读取文件 test.txt 统计其中的英文字母的大写个数和小写字母个数，并将结果输出。

5. 编写一个程序 Digital.java，随机生成 10 个数，取值范围为 0～50。要求将生成的 10 个数从小到大排序写入 n.txt 文件保存。

6. 编写一个程序 HandInput.java，接受用户的键盘输入，存入指定的文件。用户的输入以行为单位，当用户输入 end 时，程序结束。如果指定的文件已经存在，程序提示用户，并结束程序。

7. 编写一个程序，类名为 WordCount，统计单词 hello 在一篇英文文章（保存在文件 article.txt）中出现的次数，要求统计时忽略单词的大小写，统计结果在屏幕上打印出来的格式为"单词＊＊＊在文章＊＊＊中出现的次数为：10"。

提示：下面是 String 类中的几个方法。

（1）public int indexOf(String str) //返回指定子字符串在此字符串中第一次出现处的索引

（2）public int indexOf(String str,int fromIndex) //从指定的索引开始，返回指定子字符串在此字符串中第一次出现处的索引

（3）public String toUpperCase() // String 中的所有字符都转换为大写

（4）public String toLowerCase() // String 中的所有字符都转换为小写

8. 编写一个程序 RecordScore.java 用来记录某课程的成绩，要求从键盘输入学生姓名和成绩，每行输入一个学生的成绩，输入 end♯ 表示输入结束。程序要统计出课程的平均成

绩,并将输入的学生成绩和平均成绩保存到 score.txt 文件。(提示:字符串的方法 split(" ")可以返回字符串中以空格分割的字符串数组。如 str="张三 80";String [ ] s=str.split(" ");则 s[0]="张三",s[1]="80")

要求如下所示。

(1) 输入成绩的格式为姓名 成绩,例如:

张三 80
李四 90
王五 70

(2) 如果输入的格式或成绩有误,请给出提示信息。
(3) 保存写入 score.txt 文件的格式为

张三 80
李四 90
王五 70
平均成绩:80

9. 编写程序 ListWords.java 实现从键盘输入一个英文句子,统计该句子中英文单词的个数,将找出所有单词存放到一个数组中。例如:输入 He said,"That's not a good idea."则输出为

共有 8 个单词: He said that s not a good idea

注意:若有相同的单词可不处理,即两个相同的单词可分别存入数组。

10. 小张用 Java 开发了一个日历备忘录程序 Calendar.java,存于 D:/project/src 目录下,假设所有代码都是规范编写,请您编写一个 Java 程序 CodeNumber.java 统计该日历备忘录程序文件 Calendar.java 大概有多少行代码(注释的内容也可统计在内)。

11. 编写一个程序,类名为 FileMerger,将指定的多个文本文件合并到一个文件。要求类的名称为 MergeTxt。在命令行状态下,该程序的运行形式如下:

java MergeTxt 目标文件名 合并文件 1 合并文件 2 … 合并文件 n

如 java MergeTxt mergeFile.txt file1.txt file2.txt file3.txt 将 file1.txt、file2.txt、file3.txt 三个文件的内容合并后放入 mergeFile.txt 中。要求程序能够判断文件的类型是否正确,如果类型不正确或者给定的文件不存在时要给出错误信息。

12. 编写一个文件操作类 FileOperate,实现以下方法。

(1) 创建指定的目录 makeDirectory:如果指定的目录的父目录不存在则创建其目录树上所有需要的父目录。

(2) 清空指定目录中的所有文件 emptyDirectory 方法:这个方法将尽可能删除所有的文件,但是只要有一个文件没有被删除都会返回 false。

(3) listAll 方法:列出目录中的所有内容,包括其子目录中的内容。

(4) getTypePart 方法:得到指定目录下所有以某后缀命名的所有文件名。

(5) 搜索文件 SearchFile 方法:搜索给定目录下的指定文件,支持模糊查询和深度搜索。如 test.*,则返回所有以 test. 开头的文件名。

13. 编写一个简单的学生信息管理程序 StudentInfoManager.java,利用向量 Vector 记录实现学生信息,能支持学生对象的增加、删除操作,每个学生包括学号、姓名、性别,信息从键盘输入。删除学生必须输入学生的学号。可以设计一个操作菜单,包括"增加"、"删除"、"显示"、"退出"4 个选项。

14. 实现一个简单的学生管理系统程序 StudentMIS.java,能支持学生数据(Student)的显示、增加、修改和删除等操作,学生信息包括学号、姓名、性别、专业等。具体要求如下:

(1) 学生数据按对象序列化写入 student.dat 文件。

(2) 程序启动后,从 student.dat 文件读取学生对象存入 ArrayList 中。

(3) 增加学生数据时将学生对象 Student 存入 ArrayList,按保存后再写入 student.dat。

(4) 删除学生必须输入学生的学号。

(5) 修改学生必须输入学号,然后输入姓名、性别和专业进行修改。

(6) 保存时将当前 ArrayList 中的所有学生对象写入 student.dat。

(7) 退出前也要将当前 ArrayList 中的所有学生对象写入 student.dat。

(8) 设计一个操作菜单,包括"增加"、"修改"、"删除"、"显示"、"保存"、"退出"等 6 个选项。

第9章 图形用户界面编程

9.1 知 识 点

(1) AWT 与 Swing。
(2) 容器组件,包括 JFrame、JPanel、JScrollPane 和 JSplitPane。
(3) 菜单和工具条。
(4) 基本组件,包括 JLabel、JTextField、JPasswordField、JButton、JCheckBox、JRadioButton、JComboBox、JList、JTextArea、JTable 和 JProgressBar。
(5) 布局管理器,包括 FlowLayout、BorderLayout、BoxLayout、GridLayout、CardLayout 和 GridBagLayout。
(6) 事件处理模型,包括事件处理机制、事件对象、监听器接口。
(7) 鼠标事件处理,事件适配器,键盘事件处理。

重点:容器组件、基本组件、布局管理器、事件处理模型。
难点:复杂界面布局的应用,事件处理模型。

9.2 例 题

【例 9-1】 编写一个数据自增/自减的程序,允许用户在文本字段中输入一个数,每当用户单击一次自增按钮就将此数加一,用户单击一次自减按钮就将此数减一。界面效果如图 9-1 所示。

【解析】 编写图形用户界面程序基本步骤如下。
(1) 引入用到的组件和事件包或类,一般包含 java.awt.*、java.awt.event.* 和 java.swing.* 等三个包或这些包中的类。

图 9-1 数据自增/自减界面

(2) 声明类并实现相关事件的接口。
(3) 声明界面中需用到的组件。
(4) 在构造方法中创建组件的对象,对界面中的组件进行合理布局,并注册相关组件的事件监听器,最后设置界面大小并显示界面。
(5) 实现接口的方法,即事件的处理方法。
(6) 编写 main 方法进行测试。

本例中需用到一个输入框 JTextField 和两个按钮 JButton,处理按钮事件的监听器接口是 java.awt.event.ActionListener。值得注意的是,作为类成员变量定义的组件,不要在构造方法中重新定义,否则导致执行程序时在事件处理的方法中出现空指针异常(NullPointerException)。本例参考代码如下:

1.　/*引入用到组件和事件*/

```java
2.  import java.awt.*;
3.  import java.awt.event.*;
4.  import javax.swing.*;
5.  class Incrementor implements ActionListener{        //声明类并实现事件接口
6.      JTextField numberTxf;                           //本文输入框
7.      JButton incrementBtn,decrementBtn;              //自增和自减按钮
8.      /*在构造方法中创建组件的对象,对界面进行合理布局,并注册相关组件的事件监听器,
            最后设置界面大小并显示界面*/
9.      public void Incrementor (){
10.         JFrame frm=new JFrame("Incrementor");
11.         Container c=frm.getContentPane();
12.         c.setLayout(new FlowLayout());
13.         numberTxf=new JTextField("0",5);
14.         c.add(numberTxf);
15.         incrementBtn=new JButton("Increment");
16.         c.add(incrementBtn);
17.         /*注册自增按钮的事件监听器*/
18.         incrementBtn.addActionListener(this);
19.         decrementBtn=new JButton("Decrement");
20.         c.add(decrementBtn);
21.         /*注册自减按钮的事件监听器*/
22.         decrementBtn.addActionListener(this);
23.         frm.setSize(300,100);                       //设置界面大小
24.         frm.setVisible(true);                       //显示界面
25.     }
26.     /*事件的处理方法*/
27.     public void actionPerformed(ActionEvent e) {
28.         /*从输入框中获取原始值并转化为整数*/
29.         int oldNum=Integer.parseInt(numberTxf.getText());
30.         int newNum=oldNum;
31.         if(e.getSource()==incrementBtn){            //如果是自增按钮
32.             newNum++;
33.         }
34.         else if(e.getSource()==decrementBtn){       //如果是自减按钮
35.             newNum--;
36.         }
37.         numberTxf.setText(String.valueOf(newNum));
38.     }
39.     /*编写main方法进行测试*/
40.     public static void main(String args[]) {
41.         Incrementor i=new Incrementor();
42.     }
43. }
```

【例 9-2】 用组件 JComboBox 和 JCheckBox 来演示可选项目事件,界面如下图 9-2 所示。

【解析】 本例界面中使用的组件是 JComboBox 和 JCheckBox，实现的事件监听器接口都是 java.awt.event.ItemListener，参考代码如下：

图 9-2 ItemListener 测试实例程序界面

```java
1.  import java.awt.*;
2.  import java.awt.event.*;
3.  import javax.swing.*;
4.  public class ItemDemo implements ItemListener{
5.    JFrame f;
6.    JPanel p1,p2,p3;
7.    JLabel birthPlace,hobby;
8.    JComboBox place;
9.    JCheckBox hobby1,hobby2,hobby3;
10.   public ItemDemo(String title){
11.     f=new JFrame(title);
12.     p1=new JPanel();
13.     birthPlace=new JLabel("出生地:");
14.     place=new JComboBox();
15.     place.addItemListener(this);
16.     place.addItem("杭州");
17.     place.addItem("宁波");
18.     place.addItem("温州");
19.     place.addItem("绍兴");
20.     p1.add(birthPlace);
21.     p1.add(place);
22.     f.add(p1,"North");
23.     p2=new JPanel();
24.     p2.setLayout(new GridLayout(3,1));
25.     hobby=new JLabel("业余爱好:");
26.     hobby1=new JCheckBox("运动");
27.     hobby1.addItemListener(this);
28.     hobby2=new JCheckBox("旅游");
29.     hobby2.addItemListener(this);
30.     hobby3=new JCheckBox("上网");
31.     hobby3.addItemListener(this);
32.     p2.add(hobby1);
33.     p2.add(hobby2);
34.     p2.add(hobby3);
35.     p3=new JPanel();
36.     p3.add(hobby);
37.     p3.add(p2);
38.     f.add(p3,"Center");
39.     f.setBounds(300,300,300,200);
40.     f.setVisible(true);
41.   }
```

```
42.    public void itemStateChanged(ItemEvent e){
43.      if(e.getSource()==place&&e.getStateChange()==ItemEvent.SELECTED){
44.        System.out.println("您当前选择的是"+place.getSelectedItem());
45.      }
46.      else if(e.getSource()==hobby1 &&hobby1.isSelected() ){
47.        System.out.println("您的业余爱好有:"+hobby1.getLabel());
48.      }
49.      else if(e.getSource()==hobby2 &&hobby2.isSelected()){
50.        System.out.println("您的业余爱好有:"+hobby2.getLabel());
51.      }
52.      else if(e.getSource()==hobby3 &&hobby3.isSelected()){
53.        System.out.println("您的业余爱好有:"+hobby3.getLabel());
54.      }
55.    }
56.    public static void main(String args[ ]){
57.      new ItemDemo("测试 ItemListener!");
58.    }
59.  }
```

9.3 练 习 题

9.3.1 判断题

1. JFrame、JPanel、JApplet 和 JButton 四种组件都属于容器组件。（ ）
2. BorderLayout 是面板 JPanel 的默认布局管理器。（ ）
3. 用 BorderLayout 布局的容器被分成 5 个区域。（ ）
4. 一个面板(JPanel)不能被加入另一个面板(JPanel)中。（ ）
5. 菜单需要一个 JMenuBar 对象，以使它们能被添加到 JFrame。（ ）
6. 处理文本输入框和按钮事件的监听器接口可以是 ActionListener。（ ）
7. Java 为每一个监听器接口都定义了对应的适配器类。（ ）
8. 处理下拉框和复选框事件的监听器接口都是 ItemListener。（ ）
9. 处理 JButton 的事件监听器接口是 ActionListener。（ ）
10. javax.swing.JPanel 的默认布局管理器是 FlowLayout。（ ）
11. 如果一个组件注册多个监听器，事件只会被最后一个监听者处理。（ ）
12. 单击 JList 组件会产生 ActionEvent 事件，由 ActionListener 处理。（ ）
13. BorderLayout 布局管理器把组件排列在北、南、东、西和中间区域。（ ）

9.3.2 选择题

1. 下列说法哪个是正确的？（ ）
   A. BorderLayout 是面板的默认布局管理器
   B. 当鼠标指针位于一个 GUI 组件的边上时,会发生一个 MouseEvent 事件

C. 一个面板(JPanel)不能被加入到另一个面板(JPanel)中
D. 在 BorderLayout 中,添加到 NORTH 区的两个按钮将并排显示

2. FlowLayout 布局管理器按照组件加入容器的次序从(　　)到(　　)排列组件。
   A. 上、下　　　　B. 右、左　　　　C. 左、右　　　　D. 前、后

3. 在 Java 中,要处理 JButton 类对象的事件的接口是(　　)。
   A. FocusListener　　　　　　　　B. ComponentListener
   C. WindowListener　　　　　　　D. ActionListener

4. 下列哪些接口在 Java 中没有定义相对应的 Adapter 类?(　　)
   A. MouseListener　　　　　　　　B. ItemListener
   C. WindowListener　　　　　　　D. ActionListener

5. 下列哪种 Java 组件是容器组件?(　　)
   A. 列表框 JList　　　　　　　　 B. 下拉式列表框 JComboBox
   C. 面板 JPanel　　　　　　　　　D. 命令式菜单项 JMenuItem

6. 使用下面哪一个布局管理器,使得当 JFrame 的大小被改变时 JFrame 中的按钮的位置可能会被发生改变?(　　)
   A. BorderLayout　　　　　　　　B. FlowLayout
   C. CardLayout　　　　　　　　　D. GridLayout

7. 使用下面哪一个布局管理器,使得当 JFrame 的大小被改变时 JFrame 中的按钮的大小不会被改变?(　　)
   A. BorderLayout　　　　　　　　B. FlowLayout
   C. GridBagLayout　　　　　　　 D. GridLayout

8. 对 Java 中的 JComboBox 下拉式列表框进行选择其中的内容,实现该事件监听应采用以下哪种接口?(　　)
   A. ActionListener 接口　　　　　B. MouseMotionListener 接口
   C. ItemListener 接口　　　　　　D. WindowListener 接口

9. JPanel 的默认布局管理器是(　　)。
   A. BorderLayout　　　　　　　　B. FlowLayout
   C. CardLayout　　　　　　　　　D. GridLayout

10. 一个组件在水平方向会改变大小,但垂直方向的大小不变,则放到什么位置?(　　)
    A. BorderLayout 布局的 North 或 South 位置
    B. BorderLayout 布局的 East 或 West 位置
    C. BorderLayout 布局的 Center 位置
    D. GridLayout 布局中

11. 关于 Java 的事件处理机制,下面的说法正确的是(　　)。
    A. 每一个组件都可以发生任意类型的事件
    B. 事件处理机制有三个要素:事件源、监听器和处理事件的接口
    C. 监听器不必实现接口中的所有方法,只可实现需要的一个或几个方法即可
    D. 任意组件产生的事件,都可以定义事件适配器来实现

12. 下面说法错误的是（　　）。
    A. 单击 JCheckBox（复选框）产生 ItemEvent 事件，并由 ItemListener 处理
    B. 处理 JButton 和 JTextField 事件监听器接口都可以是 ActionListener
    C. 一个面板（JPanel）可以加入到另一个面板（JPanel）中
    D. 采用 BorderLayout 布局时，添加到 North 区的两个按钮将并排显示
13. 下面说法错误的是（　　）。
    A. 单击 JList（列表）可产生 ItemEvent 事件，并由 ItemListener 处理
    B. 处理鼠标移动事件的事件监听器接口是 MouseListener
    C. 处理鼠标事件的监听器接口只有 MouseListener 和 MouseMotionListener
    D. 在 JTextField 和 JPasswordField 中输入数据后按回车（Enter）键会激发一个事件，事件监听器接口都是 ActionListener
14. 下列说法哪个是正确的？（　　）
    A. 如果一个组件注册多个监听者，该组件的事件只会被最后一个注册的监听者处理
    B. 如果一个组件注册多个监听者，该组件的事件将被所有注册的监听者处理
    C. 一个组件注册多个监听者将导致编译出错
    D. 一个组件一旦被注册监听器，则无法将该监听者移去
15. JTextArea 实现了可以处理（　　）文本信息的文本框，但其不能自动进行（　　）处理。
    A. 单行，滚屏　　　　　　　　　　　　B. 多行，编辑
    C. 多行，滚屏　　　　　　　　　　　　D. 单行，编辑
16. 常规菜单是由（　　）、（　　）和菜单项 JMenuItem 组成。
    A. JMenuItem、JCheckBoxMenuItem
    B. JButton、JRadioButton
    C. JPopupMenu、JMenuItem
    D. JMenuBar、JMenu

9.3.3　程序阅读题

1. 阅读下面的程序，写出带下划线(1)~(4)的语句或注释。

```
1.  import java.awt.*;
2.  import javax.swing.*;
3.  public class Test _____(1)_____ {        //继承 JFrame
4.    public Test(){
5.      super("Concentric");                       //_____(2)_____
6.      setSize(200,100);                          //_____(3)_____
7.      setVisible(true);                          //_____(4)_____
8.    }
9.    public static void main( String args[]){
10.     Test application=new Test();
11.     application.setDefaultCloseOperation(JFrame.EXIT_ON_CLOSE);
```

12.   }
13.   }

2. 仔细阅读下面的程序,简单的画出 GUI 的界面

```java
import java.awt.*;
import javax.swing.*;
public class TestBorderLayout extends JFrame {
    public TestBorderLayout() {
        super("BorderLayout 实例");
        setLayout(new BorderLayout());
        add(new JButton("one"),BorderLayout.NORTH);
        add(new JButton("two"),BorderLayout.SOUTH);
        add(new JButton("three"),BorderLayout.WEST);
        add(new JButton("four"),BorderLayout.EAST);
        add(new JButton("five"),BorderLayout.CENTER);
    }
    public static void main(String args[]){
        TestBorderLayout t=new TestBorderLayout();
        t.pack();
        t.setVisible(true);
    }
}
```

3. 阅读下面程序代码,找出其中的错误并改正其中的错误。

```
1.    import java.awt.*;
2.    import java.awt.event.*;
3.    import javax.swing.*;
4.    public class ButtonTest extends JFrame {
5.      private JButton plainButton, fancyButton;
6.      public ButtonTest(){
7.        super( "Testing Buttons" );
8.        Container container=getContentPane();
9.        container.setLayout( new FlowLayout() );
10.       plainButton=new JButton( "Plain Button" );
11.       container.put( plainButton );
12.       fancyButton=new JButton( "Fancy Button");
13.       container.put( fancyButton );
14.       ButtonHandler handler=new ButtonHandler();
15.       fancyButton.addItemListener( handler );
16.       plainButton.addItemListener( handler );
17.       setSize( 275, 100 );
18.       setVisible( true );
19.     }
20.     public static void main( String[] args ) {
21.       ButtonTest application=new ButtonTest();
```

```
22.          application.setDefaultCloseOperation(JFrame.EXIT_ON_CLOSE );
23.     }
24.     private class ButtonHandler implements ActionListener {
25.        public void itemStateChanged( ActionEvent event ) {
26.             JOptionPane. showMessageDialog ( null, " You pressed: " + event.getActionCommand() );
27.        }
28.     }
29. }
```

4. 阅读下面的程序代码,根据注释要求补充、完成(1)~(4)代码(下划线是需要补充的地方),最后画出程序运行结果的界面。

```
1.  import java.awt.*;
2.  import _____(1)_____;
3.  public class ColorSelect extends JFrame {
4.      private JButton ok, cancel;
5.      private JCheckBox background, foreground;
6.      private JComboBox colorList;
7.      private JPanel panel, panel2;
8.      private Container c;
9.      public ColorSelect(){
10.         super( "ColorSelect" );
11.         c=_____(2)_____;                //得到框架的内容面板(容器)
12.         _____(3)_____;                  //设置容器 c 的布局为 BorderLayout
13.         colorList=new JComboBox();
14.         colorList.addItem( "RED" );
15.         c.add( colorList, BorderLayout.NORTH );
16.         panel=new JPanel();
17.         background=new JCheckBox( "Background" );
18.         foreground=new JCheckBox( "Foreground" );
19.         panel.add( background );
20.         panel.add( foreground );
21.         c.add( panel, BorderLayout.CENTER );
22.         ok=new JButton( "Ok" );
23.         cancel=new JButton( "Cancel" );
24.         panel2=new JPanel();
25.         panel2.add( ok );
26.         panel2.add( cancel );
27.         panel2.add( cancel );
28.         c.add( panel2, BorderLayout.SOUTH );
29.         setSize( 300, 125 );
30.         _____(4)_____;
31.     }
32.     public static void main ( String args[] ){
33.         ColorSelect app=new ColorSelect();
```

```
34.        app.setDefaultCloseOperation( EXIT_ON_CLOSE );
35.    }
36. }
```

5. 阅读下面的程序，编译是否出错。如没有编译错误，画出执行结果。

```
1.  import java.awt.*;
2.  import javax.swing.*;
3.  public class GridLayoutExample{
4.    public static void main( String args[] ){
5.      JFrame app=new JFrame("GridLayout");
6.      Container container=app.getContentPane();
7.      container.setLayout( new GridLayout(2, 3) );
8.      JButton [] b=new JButton[5];
9.      for (int i=0; i<5; i++){
10.        String s="Button " + (i+1);
11.        b[i]=new JButton(s);
12.        container.add( b[i] );
13.     }
14.     app.setSize( 300, 100 );
15.     app.setVisible( true );
16.     app.setDefaultCloseOperation(JFrame.EXIT_ON_CLOSE );
17.   }
18. }
```

6. 下面的程序是一个简单的图形界面程序 ButtonDemo.java，界面标题为"按钮测试"，窗口大小为 200×100 像素，使用 FlowLayout 布局，界面上有一个文本输入框 JTextField(长度为 20)和两个按钮(分别为 OK 和 Cancel)，单击 OK 按钮，则在文本输入框显示"您按了 OK 按钮!"，单击 Cancel 按钮，则在文本输入框显示"您按了 Cancel 按钮!"。根据注释要求补充、完成划线(1) ~(5)代码。

```
import java.awt.*;
import javax.swing.*;
    (1)                    ;            //加载事件处理的包
public class ButtonDemo   (2)            {
    private JButton ok,cancel;
    private JTextField tf;
    public ButtonDemo(){
        (3)                              //创建窗口界面,并设标题为"按钮测试"
        Container c=f.getContentPane();
        c.setLayout(new FlowLayout());
        tf=new JTextField(20);
        c.add(tf);
        ok=new JButton("OK");
        ok.addActionListener(this);
        cancel=  new JButton("Cancel");
```

```
            cancel.addActionListener(this);
            c.add(ok);
            c.add(cancel);
            f.setSize(200,100);
            f.setVisible(true);
        }
        public void ___(4)_____ {
            if(___(5)_____){//如果单击 OK 按钮)
                tf.setText("您单击 OK 按钮!");
            }
            else{
                tf.setText("您单击 Cancel 按钮!");
            }
        }
        public static void main(String args[ ]){
            new ButtonDemo();
        }
    }
```

7. 下面是一个数字-英文转换的图形用户界面程序,包括一个文本框和一个标签。在文本框输入一个数字(0～9),按回车键,在标签处显示对应的英文单词,即 0—zero,1—one,…,9—nine。若输入非数字字符,在标签处显示"输入出错!"。若输入的数据超过 0～9 的范围,提示"输入的数据超出范围!"。请将划线上(1)～(5)的语句补充完整。

```
import java.awt.*;
import java.awt.event.*;
import javax.swing.*;
public class EnglishToNumber extends JFrame   ___(1)___  {   //实现监听器接口
    JLabel wordInfo;
    JTextField number;
    String word[]={"zero","one","two","three","four",
                   "five","six","seven","eight","nine","ten"};
    public  EnglishToNumber (){
        super("数字英文转换器");
        number=new JTextField(10);
        wordInfo=new JLabel("英文单词");
        setLayout(new FlowLayout());
        getContentPane().add(number);
        getContentPane().add(wordInfo);
        ___(2)_____                        //注册输入框组件的事件监听器
        setSize(300,100);
        setVisible(true);
    }
    public void actionPerformed(ActionEvent e) {
        String s=number.getText();
        String info="无";
```

```
        int i=0;
        try{
            i=    (3)                         //将输入框获取的内容转化为整数
            if(i>=0&&i<10){
                info=word[i];
            }
            else{
                info="输入的数据超出范围!";
            }
        }catch(Exception e1){
                (4)          ;                //设置"输入错误"的信息
        }
            (5)              ;                //设置标签的信息
    }
    public static void main(String args[]){
        new EnglishToNumber();
    }
}
```

8. 下面的程序是文本编辑程序 TextEdit.java,用户界面中含一个 TextArea 组件和一个 Button 组件(采用 BorderLayout 布局,TextArea 组件放在 Center 区,Button 组件放在 South 区),用单击按钮,程序将 TextArea 组件中的内容写入文件 C:/mytext.txt 中。根据注释要求补充、完成划线(1)～(5)代码。

```
import java.awt.*;
import java.awt.event.*;
import javax.swing.*;
    (1)              ;                        //加载输入输出流包
public class TextEdit     (2)            {
                                              //TextEdit 类继承 JFrame,实现 ActionListener 接口
    JTextArea text;
    JButton save;
    public TextEdit(){
        text=new JTextArea();
        save=new JButton(" 保  存 ");
        Container c=getContentPane();
            (3)                      ;        //将 JTextArea 组件放在 Center 区
        add(save,BorderLayout.SOUTH);
        save.addActionListener(this);
        setSize(300,200);
        setVisible(true);
    }
    public void     (4)        {
        if(e.getSource()==save){
            try{
                PrintWriter p=new PrintWriter(   (5)   )
```

```
                    p.println(text.getText());
                    p.close();
                }catch (IOException ioe) {
                    ioe.printStackTrace();
                }
            }
        }
    }
    public static void main(String[]args){
        new TextEdit();
    }
}
```

9.3.4 编程题

1. 编写一个简单的图形界面程序 ButtonInfoShowDemo.java，界面使用 JFrame 实现，窗口大小为 300×100 像素，使用 FlowLayout 布局，有两个 JButton 按钮（名称为 b1、b2）和一个长度为 20 的文本框 JTextField，单击按钮，在 JTextField 上显示该按钮的信息。

2. 设计一个界面程序 ClickMe.java，内含一个按钮，开始运行时，按钮显示"Click Me"字样，当按下按钮时，按钮上面的文字变成"Click Me Again"，再按一次，则变回原来的"Click Me"字样，如此循环。

3. 编写一个图形界面程序 CopyText.java，包括两个文本框和一个按钮。当单击按钮时，可以把一个文本框中的内容复制到另一个文本框中。

4. 编写一个图形界面程序 LeapYear.java，包括两个文本框和一个按钮。在第一个文本框中输入一个年份。当单击按钮时可以判断出第一个文本框中输入的是否是闰年，结果显示在第二个文本框中。

5. 编写一个图形界面程序 WelcomeYou.java，包括两个文本框和一个按钮。在第一个文本框中输入一个姓名，当单击按钮时，在第二个文本框中输出"姓名 Welcome you！"。例如，输入"张三"，输出"张三 Welcome you"。

6. 编写一个图形界面程序 ClickButton.java，其功能为在界面中有一个按钮，当不断地单击按钮就显示它被单击的次数。

7. 编写一个图形界面程序 Keyboard.java，用来处理指定的键盘事件，当在键盘上按下某一个字母键时，显示该键编码和字母本身。

8. 编写一个图形界面的程序 Sum.java，获取两个文本域的输入并求和，然后显示在第三个文本域中。

9. 编写一个简单的文本编辑程序 TextEdit.java，用户界面中含一个 JTextArea 组件和一个"保存"JButton 组件（采用 BorderLayout 布局，JTextArea 组件放在 Center 区，JButton 组件放在 South 区），单击按钮，程序将 JTextArea 组件中的内容写入文件 mytext.txt 中。

10. 使用 Swing 组件编写一个支持中文文本编辑程序 ChineseTextEdit.java，要求如下。

（1）用户界面大小为 400×200 像素，如图 9-3 所示。

(2) 程序启动后,多行文本输入框 JTextArea 中显示当前目录下 myText.txt 文件中原有的内容,如果该文件不存在,则新建该文件。

(3) "保存"按钮功能:将多行文本输入框 JTextArea 中的内容写入 myText.txt 文件中保存。

(4) "取消"按钮功能:将多行文本输入框 JTextArea 中的内容清空。

(5) "退出"按钮功能:退出程序。

(6) 窗口事件不处理。

11. 编写一个用户登录程序 UserLoginApp.java,具体要求如下。

(1) 用户界面大小为 200×160 像素,如下图 9-4 所示。

图 9-3 文本编辑器

图 9-4 用户登录程序

(2) 用户类型包括学生用户和教师用户,默认为学生用户。

(3) "确定"按钮功能:如果用户名为空,则打印"用户名不可为空!";如果密码为空,则打印"密码不可为空!";如果是学生用户,用户名和密码都是 s,登录成功则打印"学生用户登录成功";如果是教师用户,用户名和密码都是 t;登录成功则打印"教师用户登录成功";如果登录不成功,则打印"用户名不存在或者密码不正确!"。

(4) "取消"按钮功能:将用户名和密码的输入框清空。

(5) "退出"按钮功能:退出程序。

12. 编写一个数字-英文转换的图形用户界面程序 EnglishToNumber.java,包括 1 个文本框和一个标签。在文本框输入一个数字(0~9),按回车键,在标签处显示对应的英文单词。0—zero,1—one,…,9—nine,如图 9-5 所示。若输入非数字字符,在标签处显示"输入错误!"。若输入的数据超过 0~9 的范围,提示"输入的数据超出范围!"。

(a) 输入 0~9 数字后的情况

(b) 输入错误的情况

(c) 输入的数字超过范围的情况

图 9-5 各种输入情况

13. 编写一个将华氏温度转换为摄氏温度的图形界面程序 TempTrans.java，界面效果如图 9-6 所示。要求从文本输入框输入华氏温度，单击"转换"按钮后，在文本框中显示转换后的摄氏温度，转换后的摄氏温度精确到 2 位。其中，华氏-摄氏温度转换公式为：摄氏温度＝5/9×华氏温度－32）。

图 9-6　华氏-摄氏温度转换器

提示：

```
java.text.DecimalFormat df=new java.text.DecimalFormat("###.##");
String s=df.format(123.4567);
System.out.println(s);
```

输出结果为：123.46。

14. 使用图形用户界面编写一个猜数游戏程序 GuessNumber.java，程序生成一个 10 以内的整数，用户从键盘输入猜测的数，程序提示猜的数比生成的数大还是小，直到猜对为止。请自己设置图形用户界面的布局。

第 10 章 多 线 程

10.1 知 识 点

(1) 进程和线程的概念,线程的生命周期。
(2) 如何编写线程程序。
(3) 线程的控制与调度。
(4) 线程同步处理。
重点:线程的创建、线程的控制与调度,多线程同步问题。
难点:多线程同步问题和处理方法。

10.2 例 题

【例 10-1】 编写程序创建三个线程,每个线程循环打印 5 次和自己的序号。

【解析】 Java 编写线程程序有两种方法。

(1) 通过实现接口 java.lang.Runnable 创建线程。首先,定义一个类实现 Runnable 接口;然后将该类的实例作为参数传给 Thread 类的一个构造方法,从而创建一个线程。

(2) 通过继承类 java.lang.Thread 创建线程。首先,通过继承 Thread 类,重写 run()方法来定义线程体;然后,创建该子类的对象创建线程。

本例的参考代码如下:

```
1.    class MultiThread extends Thread{              //继承线程类 Thread
2.      int threadNum;
3.      public static void main(String args[]){
4.        MultiThread array[]=new MultiThread [3];
5.        for (int i=0;i<3;i++)
6.          array[i]=new MultiThread (i+1);          //创建线程对象
7.        for (int i=0;i<3;i++)
8.          array[i].start();                        //启动线程
9.      }
10.     MultiThread (int SerialNum){
11.       super();
12.       threadNum=SerialNum;
13.     }
14.     public void run(){
15.       int MySerialNum=0;
16.       for(int j=1;j<=5;j++){
17.         MySerialNum++;
18.         System.out.println("<"+j +">loop:" +MySerialNum);
```

```
19.        System.out.println("thread:"+threadNum+" bye.");
20.      }
21.    }
22. }
```

多个线程运行时,调度策略为按优先级调度,级别相同时,由操作系统按时间片来分配。对本例的程序进行多次运行,发现每次结果都不一样,说明多个进程运行时执行顺序是交叉的。

【例 10-2】 用线程同步处理的方法编写程序演示生产者和消费者模型。

【解析】 使用某种资源的线程称为消费者,产生或释放这个资源的线程称为生产者。假设生产者生成 10 个整数(0～9),存储到一个共享对象中,并把它们打印出来。每生成一个数就随机休眠 0～100ms,然后重复这个过程。一旦这 10 个数可以从共享对象中得到,消费者将尽可能快地消费这 10 个数,即把它们取出后打印出来。因此,理想的生产者和消费者模型是生产者产生新数据,消费者马上能取走该数据。

一个程序的各个并发线程中对同一个对象进行访问的代码段,成为临界区。在 Java 语言中,临界区可以是一个语句块或一个方法,并且用 synchronized 关键字标识。线程的同步处理需用方法 wait 和 notify 或 notifyAll 进行处理,方法 wait 使得当前线程等待,方法 notify 或 notifyAll 则唤醒当前或所有等待的线程。本例首先创建一个共享对象的类,包含共享资源变量、获取该资源变量值的方法 get 和给该资源变量赋值的方法 put,其中 get 和 put 方法需用 synchronized 修饰,并对共享的资源变量进行数据对象锁的同步处理。然后定义 Producer 和 Consumer 分别代表生产者和消费者线程,最后编写一个测试类模拟生产者和消费者模型。本例的参考代码如下。

(1) 共享对象的程序代码。

```
1.  public class Share {
2.    private int contents;
3.    private boolean available=false;
4.    public synchronized int get() {
5.      while (available==false) {
6.        try {
7.          wait();              //线程等待并暂时释放共享数据对象的锁
8.        }
9.        catch (InterruptedException e) { }
10.     }
11.     available=false;
12.     notifyAll();              //释放对象锁,通知正在等待的线程重新占有锁并运行
13.     return contents;
14.   }
15.   public synchronized void put(int value) {
16.     while (available==true) {
17.       try {
18.         wait();
19.       }
```

```
20.        catch (InterruptedException e) { }
21.      }
22.      contents=value;
23.      available=true;
24.      notifyAll();
25.    }
26. }
```

(2) 模拟生产者的程序代码。

```
1.  public class Producer extends Thread {
2.    private Share shared;
3.    private int number;
4.    public Producer(Share s, int number) {
5.      shared=s;
6.      this.number=number;
7.    }
8.    public void run() {
9.      for (int i=0; i<10; i++) {
10.       shared.put(i);
11.       System.out.println("生产者"+this.number+"输出的数据为:"+i);
12.       try {
13.         sleep((int)(Math.random() * 100));
14.       }
15.       catch (InterruptedException e) { }
16.     }
17.   }
18. }
```

(3) 模型消费者的程序代码。

```
1.  public class Consumer extends Thread {
2.    private Share shared;
3.    private int number;
4.    public Consumer(Share s, int number) {
5.      shared=s;
6.      this.number=number;
7.    }
8.    public void run() {
9.      int value=0;
10.     for (int i=0; i<10; i++) {
11.       value=shared.get();
12.       System.out.println("消费者"+this.number+"得到的数据为:"+value);
13.     }
14.   }
15. }
```

(4) 测试主程序的代码。

```
1.  public class PCTest {
2.      public static void main(String[] args) {
3.          Share s=new Share();
4.          Producer p=new Producer(s,1);
5.          Consumer c=new Consumer(s,1);
6.          p.start();
7.          c.start();
8.      }
9.  }
```

10.3 练 习 题

10.3.1 判断题

1. 线程可以用 yield 使同优先级的线程先执行。　　　　　　　　　　　(　　)
2. 线程 t1 中执行 t2.sleep(5000)语句,则线程 t2 休眠 5s。　　　　　　　(　　)
3. 线程 t 运行就是运行其 run()方法,所以 t.start()等价于 t.run()。　　　(　　)
4. 如果有高优先级的线程就绪,则正在执行的低优先级线程将暂停执行。(　　)
5. 线程运行中调用 sleep 方法进入阻塞状态,sleep 结束后线程马上处于运行
(running)的状态。　　　　　　　　　　　　　　　　　　　　　　(　　)
6. 多个线程的运行顺序一定是按线程启动的顺序进行的。　　　　　　　(　　)
7. run 方法是运行线程的主体,若 run 方法运行结束,线程就消亡了。　　(　　)
8. Java 线程设计中,notify()方法会激活在等待集中的所有线程。　　　　(　　)
9. 每个 Java 线程的优先级都设置在常数 1～12,默认的优先级设置为常数 6。(　　)
10. 一个线程对象的具体操作是由 run()方法的内容确定的,但是 Thread 类的 run()方法是空的,其中没有内容;所以用户的线程程序要么派生一个 Thread 的子类并在子类里重新定义 run() 方法,要么使一个类实现 Runnable 接口并覆盖 run()方法。(　　)
11. 语句 mt.setPriority(10)将线程对象 mt 的优先级设置为 10。　　　　　(　　)
12. 某程序中的主类不是 Thread 的子类,也没有实现 Runnable 接口,则这个主类运行时不能控制其他线程睡眠。　　　　　　　　　　　　　　　　　　　(　　)
13. sleep()和 wait()方法都使当前运行线程放弃处理器和它所占用的同步资源管理。
　　　　　　　　　　　　　　　　　　　　　　　　　　　　　　　(　　)
14. 挂起、阻塞或等待的线程都能够恢复运行,但是停止运行的线程将不可能再复生。
　　　　　　　　　　　　　　　　　　　　　　　　　　　　　　　(　　)
15. 线程使用 sleep()方法休眠后,可以用 notify()方法唤醒。　　　　　　(　　)

10.3.2 选择题

1. 在 java 程序中,下列关于线程的说法错误的是(　　)。
   A. run 方法是运行线程的主体

B. 多个线程运行时执行顺序是按顺序执行的

C. 如果 run 方法执行结束了,说明线程死亡了

D. 在 Java 中,高优先级的可运行线程会抢占低优先级线程

2. 运行下列程序,会产生什么结果?(   )

```
1.  public class X extends Thread implements Runnable{
2.      public void run(){
3.          System.out.println("this is run()");
4.      }
5.      public static void main(String args[]){
6.          Thread t=new Thread(new X());
7.          t.start();
8.      }
9.  }
```

A. 第 1 行会产生编译错误

B. 第 6 行会产生编译错误

C. 第 6 行会产生运行错误

D. 程序没有编译错误,可正常运行

3. 编译、运行下列程序,会产生什么结果?(   )

```
1.  public class TestRunnable implements Runnable {
2.      public void run(){
3.          System.out.print(1);
4.          try{
5.              Thread.sleep(1000);
6.          }
7.          catch(Exception e){ }
8.          System.out.print(2);
9.      }
10.     public static void main(String args[]){
11.         Thread t=new Thread(new TestRunnable());
12.         t.start();
13.     }
14. }
```

A. 程序无法通过编译

B. 程序可以通过编译并正常运行,结果输出 1

C. 程序可以通过编译,结果输出 12

D. 程序可以通过编译,但运行时会抛出异常

4. 下面哪个方法可以在任何时候被任何线程调用?(   )

A. notify()   B. wait()

C. notifyAll()   D. sleep()

5. 下面哪个方法是实现 Runnable 接口所必需的?(   )

A. wait()   B. run()

C. sleep() D. notify()

6. Thread 类用来创建和控制线程,启动一个线程应该使用下面的哪个方法?(　　)
   A. init() B. start()
   C. run() D. notifyAll()

7. 下面关于 Java 中线程的说法不正确的是(　　)。
   A. 调用 join()方法可能抛出异常 InterruptedException
   B. sleep()方法是 Thread 类的静态方法
   C. 调用 Thread 类的 sleep()方法可终止一个线程对象
   D. 线程启动后执行的代码放在其 run 方法中

8. 给 Java 线程设定优先级的成员方法是(　　)。
   A. getPriority() B. setPriority()
   C. getTread() D. setTread()

9. 下列哪个情况可以终止当前线程的运行而无法恢复?(　　)
   A. 抛出一个异常对象时
   B. 当该线程调用 sleep()方法时
   C. 当创建一个新线程时
   D. 当一个优先级高的线程进入就绪状态时

10. 以下关于线程的说法错误的是(　　)。
    A. Thread 类的构造方法以实现 Runnable 接口的类的对象作为参数来创建线程
    B. 线程是按优先级调度执行的,高优先级的线程会被优先执行
    C. 多线程同步处理的目的是为了让多个线程协调地并发工作
    D. 多个线程并发执行时,线程的相对执行顺序是按线程启动的顺序来执行的

11. 某线程调用 sleep 方法,休眠结束后,将进入(　　)状态。
    A. Blocked（阻塞） B. Runnable（可运行或就绪）
    C. Running（运行） D. Dead（消亡）

12. 若编译和运行下列代码,输出结果为(　　)。

```
public class MyThread extends Thread {
  String myString="Yes";
  public void run(){
    this.myString="No";
  }
  public static void main(String[] args){
    MyThread t1=new MyThread();
    MyThread t2=new MyThread();
    t1.start();
    t2.start();
    System.out.print(t1.myString);
    System.out.print(t2.myString);
  }
}
```

}
  A. Yes No   B. Yes Yes
  C. No No   D. 无法确定

13. 关于线程的调度,下列说法哪个是正确的?( )
  A. 线程的调度采用随机原则,先占的线程优先执行
  B. 线程的调度采用占先原则,优先级低的线程优先执行
  C. 线程的调度采用占先原则,优先级高的线程优先执行
  D. 线程的调度采用随机原则,优先级为5的可先于优先级为10的线程执行

14. 以下关于线程的说法错误的是( )。
  A. Thread 的构造方法以实现 Runnable 接口的类的对象作为参数可以创建线程。
  B. 线程运行中调用 sleep 方法进入阻塞状态,sleep 结束后线程马上处于运行(Running)的状态
  C. 线程的调度采用占先原则,优先级高的线程优先执行,每个 Java 线程的优先级都设置在常数 1~10,默认的优先级设置为常数 5
  D. 当执行到同步语句 synchronized(引用类型表达式)的语句块时,引用类型表达式所指向的对象就会被锁住,不允许其他线程对其进行访问,即当前的线程独占该对象

10.3.3 程序阅读题

1. 仔细阅读下面的程序代码,请写出该程序的功能。

```java
public class Timer extends Thread{
    int time=0;
    public Timer(int time) {
        this.time=time;
    }
    public void run(){
        try{
            for(int i=1;i<=time;i++){
                Thread.sleep(1000);
                System.out.print(i+" ");
            }
        }
        catch(Exception e) {
            System.out.println(e.toString());
        }
    }
    public static void main(String args[]){
        Timer timer=new Timer(5);
        timer.start();
    }
}
```

2. 下面的程序利用线程输出从 a 到 z 的 26 个字母,每隔 1s 输出一个字母,程序不完整,请阅读程序代码,根据注释要求填写划线(1)~(5)的代码。

```
public class Test __(1)_____ {
    char charArray[]=new char[26];
    public Test(){
        for(int i=0; i<charArray.length; i++){
            charArray[i]= __(2)_____ ;
        }
    }
    public void run() {
        try {
            for (int i=0; i<charArray.length; i++) {
                __(3)_____                    //休眠 1s
                System.out.print(charArray[i]);
            }
        }
        catch (InterruptedException e) {
            e.printStackTrace();
        }
    }
    public static void main(String args[]) {
        Thread t= __(4)_____ ;                //创建线程对象
        __(5)_____ ;                          //启动线程
    }
}
```

3. 下面是一个多线程程序。请根据提示,将划线上(1)~(5)的语句补充完整。

```
class TestMultiThreads{
    public static void main(String [] args){
        Thread t1= __(1)_____ ;                //实例化线程
        __(2)_____ ;                           //将线程 t1 改名为"线程 1"
        __(3)_____ ;                           //启动线程 t1
        System.out.println ("main 方法结束。");
    }
}
class T1 __(4)_____ {                          //定义 T1 为线程体(包含线程的数据和代码)
    private int i;
    public void run(){
        while(true){
            System.out.println(i++);
            try{
                __(5)_____ ;                   //随机休眠 0~5s
            } catch(InterruptedException e) { }
        }
    }
```

        }
    }

4. 下面的程序是一个模拟龟兔赛跑的多线程程序，请将划线上(1)～(5)的语句补充完整。

```
class Animal    (1)            {              //继承线程类
    int speed;                                  //速度
    public Animal(   (2)              ) {
        super(str);                             //线程名用动物名代表
        this.speed=speed;
    }
    public void run() {
        int distance=0;
        int sleepTime;
        while (distance<=1000) {
            System.out.println(getName()+"is at"+distance);
            try {
                distance+=speed;
                sleepTime=(int)( speed+Math.random() * speed);
                sleep(sleepTime);
            } catch (InterruptedException e) {  }
        }
    }
}
public class Race {
    public static void main(String arg[]) {
           (3)          ;            //创建兔子的对象,速度为100,对象引用名为a1
           (4)          ;            //创建乌龟的对象,速度为20,对象引用名为a2
           (5)          ;            //让乌龟的运行优先级更高
        a1.start();
        a2.start();
    }
}
```

5. 下面的代码是模拟生产者-消费者问题中的共享对象对应的类 Share 的代码。当生产者线程把一个数据放入 Share 对象的 contents 中后，将 available 赋值为 true，表示生产者生产好了产品，通知消费者线程可以取走，如果消费者还没有取走产品，则生产者线程必须等待；如果消费者线程可以取走了产品，则将 available 赋值为 false，然后通知生产者继续生产，否则消费者线程必须等待。请补充完成划线(1)～(5)部分的代码，编号相同的空代码是一样的。

```
public class Share {
    private int contents;
    private boolean available=false;
```

```
    public    (1)    int get() {              //同步 get 方法
        while (    (2)    ) {                  //没有可提供的数据
            try {
                    (3)    ;                   //线程等待并暂时释放共享数据对象的锁
            } catch (InterruptedException e) {}
        }
        available=false;
            (4)    ;                           //释放对象锁,通知正在等待的线程重新占有锁并运行
        return contents;
    }
    public    (1)    void put(int value) {   //同步 put 方法
        while (    (5)    ) {                  //有可提供的数据
            try {
                    (3)    ;                   //线程等待并暂时释放共享数据对象的锁
            } catch (InterruptedException e) { }
        }
        contents=value;
        available=true;
            (4)    ;                           //释放对象锁,通知正在等待的线程重新占有锁并运行
    }
}
```

6. 阅读下面的程序,修改程序中错误的地方。

```
1.  public class Test implements Runnable {
2.      String str[]=new String[10];
3.      for (int i=0; i<str.length(); i++) {
4.          str[i]=i+"-";
5.      }
6.      public void run() {
7.          try {
8.              for (int i=0; i<str.length(); i++) {
9.                  sleep(1000);
10.                 System.out.print(str[i]);
11.             }
12.         } catch (InterruptedException e) {
13.             e.printStackTrace();
14.         }
15.     }
16.     public static void main(String args[]) {
17.         Test t=new Test();
18.         t.run();
19.     }
20. }
```

7. 阅读下面的程序，找出其中的错误，并写出最终显示的结果。

```
1.  public class A{
2.    public static void main(String args[]) {
3.      B  b=new B();
4.      b.start();
5.    }
6.  }
7.  class B implements Runnable {
8.    public void run(){
9.      for(int i=1;i<3;i++){
10.       System.out.println("第"+i+"次"+"   ");
11.       try{
12.         sleep(1000);
13.       }
14.       catch(InterruptedException e){
15.         e.printStackTrace();
16.       }
17.     }
18.    }
19.  }
```

8. 若编译和运行下列代码，输出结果是什么？

```
1.  public class MyAdd extends Thread{
2.    static int total=10;
3.    int n;
4.    public MyAdd(int n){
5.      this.n=n;
6.    }
7.    public void run() {
8.      try{
9.        sleep(n);
10.       total=total+n;
11.       System.out.println(total);
12.     }
13.     catch(Exception e){
14.       System.out.println("EXCEPTION!");
15.     }
16.   }
17.   public static void main(String[] args)  {
18.     MyAdd t1=new MyAdd(3000);
19.     MyAdd t2=new MyAdd(1000);
20.     t1.start();
21.     t2.start();
22.   }
```

23.   }

10.3.4 编程题

1. 应用 Java 中线程的概念编写一个 Java 程序(包括一个测试线程程序类 TestThread,一个 Thread 类的子类 PrintThread)。在测试程序中用子类 PrintThread 创建 2 个线程,使得其中一个线程运行时打印 5 次"线程 1 正在运行",另一个线程运行时打印 5 次"线程 2 正在运行"。

2. 利用多线程编写一个求解某范围素数的程序 FindPrime.java,每个线程负责 1000 范围:线程 1 找 1~1000;线程 2 找 1001~2000;线程 3 找 2001~3000。编写程序将每个线程找到的素数及时打印。

3. 假设一个银行的 ATM 机,它可以允许用户存款也可以取款。现在一个账户上有存款 200 元,用户 A 和用户 B 都拥有在这个账户上存款和取款的权利。用户 A 将存入 100 元,而用户 B 将取出 50 元,那么最后账户的存款应是 250 元。实际操作过程如下。

(1) 先进行 A 的存款操作:
① 得到账户的存款数额 200,耗时 2s。
② 将账户数额增加 100,耗时忽略不计。
③ 将新生成的账户结果 300 返回到 ATM 机的服务器上,耗时 2s。

(2) 再进行 B 的取款操作:
① 得到增加后账户存款数额 300,耗时 2s。
② 判断取款额是否小于账户余额,若是,则将账户数额减少 50,否则抛出异常信息,耗时忽略不计。
③ 将新生成的账户结果 250 返回到 ATM 机的服务器上,耗时 2s。

请根据以上要求,将 A 的操作和 B 的操作分别用线程 BankThread 来表示,编写一个 Java 程序 TestBankThread.java 实现该功能。

4. 请编写一个类,类名为 SubThread,是 Thread 类的子类。该类中定义了含一个字符串参数的构造方法和 run()方法,方法中有一个 for 循环,循环一共进行 5 次,循环体先在命令行显示该线程循环了第几次,然后随机休眠小于 1s 的时间,循环结束后显示线程结束信息:线程名+finished。编写一个测试程序 TestSubThrea.java,并在其中创建 SubThrea 类的三个线程对象 T1、T2、T3,它们的名称分别为 Frist、Second、Third,并启动这三个线程。

5. 编写一个堆栈类 TestStack,栈的大小为 10 个元素,每个元素可以放一个字符,实现相应的方法。再编写两个线程,线程 PushChar 包含一个 TestStack 对象,其线程方法调用 TestStack 的入栈方法,往 TestStack 对象中压入随机产生的字符;线程 PopChar 调用 TestStack 的出栈方法,取出压入的字符。要求入栈方法和出栈方法是多线程安全的,即是同步的。要求实现两个线程的交互,PushChar 做入栈操作时如果发现栈满,则等待 PopChar 出栈;PopChar 出栈操作时发现栈空,则等待 PushChar 入栈。

6*. 编写一个 Java 实现的图片查看器程序 PictureView.java,实现界面左右分割,左边显示目录和文件的名称,右边显示图片(用滚动窗口 JScrollPane 来实现图片的查看),加载图片目录时用文件选择器 JFileChooser 来实现,并具有自动翻看功能(可通过线程实现)。

界面如图 10-1 所示。

图 10-1　Java 实现的图片查看器界面

第 11 章 网 络 编 程

11.1 知 识 点

(1) 网络编程相关的基本概念。
(2) URL 编程。
(3) Socket 编程,包括基于 TCP 和 UDP 的网络程序设计。
(4) IntelAddress 类。
重点:URL 编程,Socket 编程,IntelAddress 类。
难点:基于 TCP 和 UDP 的网络程序设计。

11.2 例 题

【例 11-1】 用 URL 读取 WWW 网络资源的实例。
【解析】 用 URL 读取 WWW 网络资源数据的一般步骤如下。
(1) 创建类 URL 的实力对象,使其指向给定的网络资源。
(2) 通过类 URL 的成员方法 openStream 建立 URL 连接,并返回输入流对象的引用,以便读取数据。
(3) 通过 java.io.BufferedReader 封装输入流。
(4) 读取数据,并进行数据处理。
(5) 关闭数据流。
本例读取浙江工业大学首页(地址为 www.zjut.edu.cn)的浏览器端 HTML 文件,参考代码如下:

```
1.   import java.io.*;
2.   import java.net.*;
3.   public class URLReader {
4.   public static void main(String[] args) throws Exception {
5.     URL tirc=new URL("http://www.zjut.edu.cn);
6.     BufferedReader in=new BufferedReader(new InputStreamReader(tirc.openStream()));
7.     String inputLine;
8.     File outfile=new File("test.html");
9.     BufferedWriter out=new BufferedWriter(new FileWriter(outfile));
10.    while ((inputLine=in.readLine()) !=null) {
11.      /*从输入流中不断读取数据直到读完为止*/
12.      out.write(inputLine);            //把读入的数据写入 test.html
13.      out.newLine();
14.    }
```

```
15.        in.close();                    //关闭输入流
16.        out.close();
17.    }
18. }
```

【例11-2】 采用基于 TCP 实现的支持多客户端的 C/S 程序实例。

【解析】 基于 TCP 实现的网络程序采用 Socket 进行客户端与服务器端通信,服务器端采用类 java.net.ServerSocket 的实例使服务器能够检测到指定端口的信息,用 accept 方法等待客户连接,该方法将阻塞当前系统服务进程,直到有客户端连接成功,返回一个套接字类 java.net.Socket 对象。通过该套接字对象,新的服务子线程与连接的客户端进行通信。客户端试着创建类 java.net.Socket 的实例来请求与服务器端建立连接。本例可利用多线程实现多客户端机制。服务器端总是在指定的端口上监听是否有客户请求,一旦监听到客户端请求,服务器端就会启动一个专门的服务线程来响应该客户的请求,而服务器端本身在启动完线程之后马上又进入监听状态,等待下一个客户端的申请连接。因此,本例包括服务器端守护进程、服务器端线程程序和客户端程序。

(1) 服务器端守护进程。

```
1.  import java.io.*;
2.  import java.net.*;
3.  public class MultiTalkServer{
4.    static int clientnum=0;                //静态成员变量,记录当前客户的个数
5.    public static void main(String args[]) throws IOException {
6.      ServerSocket serverSocket=null;
7.      boolean listening=true;
8.      try{
9.        /*创建一个 ServerSocket 在端口 4700 监听客户请求*/
10.       serverSocket=new ServerSocket(4700);
11.       System.out.println("Server is running…");
12.      }catch(IOException e) {
13.       System.out.println("Could not listen on port:4700.");
14.       System.exit(-1);                   //退出
15.      }
16.      while(listening){                   //永远循环监听
17.       /*监听到客户请求,根据得到的 Socket 对象和客户计数创建服务线程,并启动之*/
18.       new ServerThread(serverSocket.accept(),clientnum).start();
19.       System.out.println("客户"+(clientnum+1)+" is connected…");
20.       clientnum++;                       //增加客户计数
21.      }
22.      serverSocket.close();               //关闭 ServerSocket
23.    }
24. }
```

(2) 服务器端线程程序。

```
1.  import java.io.BufferedReader;
```

```java
2.   import java.io.InputStreamReader;
3.   import java.io.PrintWriter;
4.   import java.net.*;
5.   public class ServerThread extends Thread{
6.     Socket socket=null;                              //保存与本线程相关的Socket对象
7.     int clientnum;                                   //保存本进程的客户计数
8.     public ServerThread(Socket socket,int num) {     //构造函数
9.       this.socket=socket;                            //初始化socket变量
10.      clientnum=num+1;                               //初始化clientnum变量
11.    }
12.    public void run() {                              //线程主体
13.      try{
14.        String line;
15.        /*由Socket对象得到输入流,并构造相应的BufferedReader对象*/
16.        BufferedReader is= new BufferedReader(new InputStreamReader(socket.getInputStream()));
17.        /*由Socket对象得到输出流,并构造PrintWriter对象*/
18.        PrintWriter os=new PrintWriter(socket.getOutputStream());
19.        BufferedReader sin= new BufferedReader(new InputStreamReader(System.in));
20.        /*标准输出上打印从客户端读入的字符串*/
21.        System.out.println("Client-"+clientnum +":" +is.readLine());
22.        line=sin.readLine();                         //从标准输入读入一字符串
23.        while (!line.equals("bye")){                 //如果该字符串为 "bye",则退出
24.          os.println(line);                          //向客户端输出该字符串
25.          os.flush();                                //刷新输出流,使客户端马上收到该字符串
26.          /*系统标准输出上打印该字符串*/
27.          System.out.println("Server:"+line);
28.          /*Client读入一字符串,并打印到标准输出上*/
29.          System.out.println("客户"+clientnum +":" +is.readLine());
30.          line=sin.readLine();                       //从系统标准输入读入一字符串
31.        } //继续循环
32.        os.close();                                  //关闭Socket输出流
33.        is.close();                                  //关闭Socket输入流
34.        socket.close();                              //关闭Socket
35.      }catch(Exception e){
36.        System.out.println("Error:"+e);
37.      }
38.    }
39.  }
```

(3) 客户端程序。

```java
1.   import java.io.BufferedReader;
2.   import java.io.InputStreamReader;
3.   import java.io.PrintWriter;
```

```
4.    import java.net.*;
5.    public class TalkClient {
6.      public static void main(String args[]) {
7.        try{
8.          /*本机的4700端口发出客户请求*/
9.          Socket socket=new Socket("127.0.0.1",4700);
10.         System.out.println("已连接服务器,请发消息给服务器处理…");
11.         /*系统标准输入设备构造BufferedReader对象*/
12.         BufferedReader sin= new BufferedReader (new InputStreamReader (System.in));
13.         /*Socket对象得到输出流,并构造PrintWriter对象*/
14.         PrintWriter os=new PrintWriter(socket.getOutputStream());
15.         /*Socket对象得到输入流,并构造相应的BufferedReader对象*/
16.          BufferedReader is= new BufferedReader (new InputStreamReader (socket.getInputStream()));
17.         String readline;
18.         readline=sin.readLine();            //从系统标准输入读入一字符串
19.         while(!readline.equals("bye")){
                                              //若从标准输入读入的字符串为 "bye"则停止循环
20.           /*从系统标准输入读入的字符串输出到Server*/
21.           os.println(readline);
22.           os.flush();                      //刷新输出流,使服务器端马上收到该字符串
23.           /*在系统标准输出上打印读入的字符串*/
24.           System.out.println("Client:"+readline);
25.           /*从Server读入一字符串,并打印到标准输出上*/
26.           System.out.println("Server:"+is.readLine());
27.           readline=sin.readLine();         //从系统标准输入读入一字符串
28.         } //继续循环
29.         os.close();                        //关闭Socket输出流
30.         is.close();                        //关闭Socket输入流
31.         socket.close();                    //关闭Socket
32.         }catch(Exception e) {
33.           System.out.println("Error"+e);   //出错,则打印出错信息
34.        }
35.      }
36.    }
```

11.3 练 习 题

11.3.1 判断题

1. java.net.SeverSocket 的成员方法 accept()是用来监听客户端的连接的,执行到该方法时,程序将被阻塞,直到监听到连接再继续执行下去。   (   )

2. 端口并不是机器上一个物理上存在的场所,而是一种软件抽象。   (   )

3. 当Socket连接对象的数据传输结束后,服务器执行Socket的shutdown方法来关闭连接。   （   ）

4. 在编写基于TCP的网络程序时,在服务器端的Socket实际上就是类java.net.ServerSocket的实例对象。   （   ）

5. 基于TCP的C/S模式网络程序至少需要两台计算机才能完成服务器与客户端的数据通信。   （   ）

11.3.2 选择题

1. 下列说法错误的是(   )。
   A. TCP是面向连接的协议,而UDP是无连接的协议
   B. 数据报传输是可靠的,可以保证包按顺序到达
   C. URL代表的统一资源定位符一共包括五个部分
   D. Socket和ServerSocket分别表示连接的Client端和Server端

2. 在Java中,有关网络通信的类和接口都集中在哪一个包中?(   )
   A. java.network   B. java.socket
   C. java.net   D. java.message

3. 下列语句不合法的是(   )。
   A. ServerSocket svrsoc=new ServerSocket(8000);
   B. URL urlBase=new URL("http://www.zjut.edu.cn");
   C. File inFile=new File("C://test.txt");
   D. BufferedReader br=new BufferedReader("C://test.txt");

4. ServerSocket构造方法的第(   )个参数指定了可以等待连接到服务器的最多客户机数目。
   A. 1   B. 2   C. 3   D. 4

5. 资源定位器对应于Java的类java.net.URL,通过其成员方法(   )获得端口。
   A. getHost()   B. getProtocol()
   C. getPort()   D. getRef()

6. 下面哪个方法是类java.net.URL的成员方法,可以打开当前URL的连接并返回输入流?(   )
   A. openConnection   B. openStream
   C. getStream   D. getConnection

7. 下面哪个方法是类java.net.Socket的成员方法,用来获取Socket的输入流?(   )
   A. getChanel   B. getConnection
   C. getInputStream   D. getStream

8. 下面哪个方法是类java.netInetAddress的静态方法,可以根据主机名创建该类的实例对象?(   )
   A. getHostName   B. getByName
   C. getHostAddress   D. getInetAddress

9. 下面哪个说法是正确的？（　　）
   A. 类java.netDatagramSocket的成员方法receive和send可以接收和发送由java.netDatagramPacket封装的数据包
   B. 类java.netDatagramSocket的成员方法receive和send可以直接接收和发送字符串实例对象
   C. 类java.netDatagramSocket可实现基于TCP协议发送和接收数据
   D. 基于TCP的网络程序在数据通信时是通过输入流（InputStream）和输出流（OutputStream）进行的，所以这类程序的数据通信只能基于字节而不能基于字符

10. 使用UDP套接字通信时，下列哪个方法用于接收数据？（　　）
    A. read()　　　　B. receive()　　　　C. accept()　　　　D. getData()

11.3.3 程序阅读题

1. 阅读下面代码片段，根据注释要求补充、完成划线(1)~(5)的代码。

```
import java.io.*;
import java.net.*;
public class SocketExample {
  public static void main(String[] args) {
    ServerSocket server;
    Socket socket;
    ObjectOutputStream output;
    ObjectInputStream input;
    try{
      //创建一个端口号为6000,可有50个客户同时连接的服务器流套接字
            (1)
      socket=server.accept();
      //创建一个Object输出流用来向客户发送数据
            (2)
      output.flush();
      //创建一个Object输入流用来接收客户发送的数据
            (3)
      //关闭输出输入流
            (4)
            (5)
      socket.close();
    }
    catch(IOException ioException) {
      System.out.println("捕获一个I/O异常");
    }
  }
}
```

2. 以下代码是基于 TCP 协议编写的一个服务器端程序,端口为 5000,要求可以同时连接多个客户端,并且每个客户端在休眠 10s 之后退出连接。请将下面的程序划线(1)~(8)部分代码补齐。

```java
import java.net.*;
import java.io.*;
public class ServerExample implements ____(1)____ {
    private Socket socket;
    private int  id;
    public ____(2)____ (Socket s, int iD.{
        socket=s;
        id=id;
    }
    public void ____(3)____ {
        try {
            Thread.sleep(10000);
            System.out.println("Socket["+id+"] is closing.");
            socket.close();
        }
        catch (Exception e) { }
    }
    public static void main(String args[]){
        int n=1;
        ____(4)____ server=null;
        try{
            server=new ____(5)____ (5000);
            System.out.println( "Server start." );
        }
        catch (IOException e) { }
        while (true){
            try{
                System.out.println("Wait NO. "+n+" connection." );
                ____(6)____ s=server.accept();
                ServerExample t=new ServerExample(s, n++);
                ____(7)____ th=new Thread(__(8)__);
                th.start();
            }
            catch (IOException e){ }
        }
    }
}
```

3. 以下代码是基于 UDP 通信的服务器端程序,请补充完成带划线(1)~(5)部分的代码。

```java
import java.net.DatagramPacket;
```

```java
import java.net.DatagramSocket;
import java.net.InetAddress;
import java.util.Date;
public class UdpServer{
  public static void main(String args[ ]){
    DatagramSocket    dSocket;
    DatagramPacket    inPacket;
    DatagramPacket    outPacket;
    InetAddress       cAddr;
    int               cPort;
    byte[ ]           inBuffer=new byte[512];
    byte[ ]           outBuffer;
    String            s;
    try{
      dSocket=   (1)   ;            //创建 DatagramSocket 对象,并指定端口为 8000
      while (true){
        inPacket=new DatagramPacket(inBuffer, inBuffer.length);
           (2)      ;               //接收数据报
        cAddr=inPacket.getAddress();
        cPort=inPacket.getPort();
        s=   (3)     ;              //获取客户端发过来的数据报中的信息
        System.out.println("接收到客户端信息:" +s);
        System.out.println("客户端主机名为:"+cAddr.getHostName());
        System.out.println("客户端端口为:" +cPort);
        Date d=new Date();
        outBuffer=d.toString().getBytes();
        outPacket=   (4)   ;        //创建给客户端发送信息的 DatagramPacket 对象
           (5)      ;               //发送数据报
      }
    }
    catch (Exception e){
      System.err.println("发生异常:" +e);
      e.printStackTrace();
    }
  }
}
```

11.3.4 编程题

1. 编写一个程序 ScanPort.java,扫描本机的小于 1000 的端口,显示正在使用的端口号,要求采用多线程方式实现。

提示:可以采用 ServerSocket 连接来判断端口是否在使用。

2. 用 Java 编写一个服务器端的程序 Server.java,该程序在 8002 端口监听客户端的请求,如果与客户端连接后,收到客户端发送的数据不是字符串"bye",则在服务器端打印客户端发来的数据,并向客户端回送一条从键盘输入的信息,若客户端发来的数据是字符串

"bye",则关闭服务器端程序。

3. 编写一个服务器端 ServerDemo.java 程序,它能够在 8003 端口响应客户的请求。如果这个请求的内容是字符串"hello"的话,服务器仅将"welcome!"字符串返回给用户。否则将用户的话追加到当前目录的文本文件 log.txt 中,并向用户返回"OK!"。

4. 使用 Socket 编写一个服务器端程序 ServerResponseHello.java,服务器端程序在端口 8004 监听,如果它接到客户端发来的"hello"请求时会回应一个"hello",对客户端的其他请求不响应。

第 12 章　数据库编程

12.1　知　识　点

（1）JDBC 模型和驱动方式。
（2）JDBC API。
（3）JDBC 编程。

重点：JDBC 驱动方式，JDBC API 和 JDBC 编程。

难点：JDBC 编程步骤。

12.2　例　　题

【例 12-1】　编写一个用户登录程序，用户类型包括教师用户和学生用户，默认为教师用户，用户的用户名和密码都在数据库中。若登录成功，则弹出"登录成功"的对话框；若登录不成功，则弹出"用户名不存在或者密码不正确！"的对话框。登录界面如图 12-1 所示。

【解析】　本例学生和教师信息都存在数据库中，假设在 mySQL 数据库 UserInfoSys 中，用户信息表 user 中的字段定义如表 12-1 所示。

该表中学生和教师用户各有 3 条记录，如表 12-2 所示。

图 12-1　用户登录界面

表 12-1　表字段

Field Name	Datatype	Len	Default	PK?	Not Null?	Auto Incr?	Comment
loginname	varchar	50	''	✓	✓		登录名
password	varchar	50					密码
usertype	int	4	0				用户类型
username	varchar	50					用户姓名

表 12-2　表中的记录

loginname	password	usertype	username
s1	s1	1	马同学
s2	s2	1	王同学
s3	s3	1	陈同学
t1	t1	0	赵老师
t2	t2	0	章老师
t3	t3	0	李老师

实现用户登录程序包括连接数据库的类 LinkDB、用户登录的数据库访问类。
UserLoginDAO 和用户登录界面的类 UserLogin，参考代码如下。

（1）连接数据库的类 LinkDB。

```
import java.sql.*;
public class LinkDB {
```

```java
    private String driverName="com.mysql.jdbc.Driver";
    private String dbUrl="jdbc:mysql:            //localhost:3306/UserInfoSys";
    private String dbUser="root";                //mySQL 数据库用户名 root
    private String dbPwd="111111";               //mySQL 数据库用户 root 的登录密码
    private Connection con;
    /*初始化,指定驱动引擎*/
    public LinkDB(){
        try{
            Class.forName(driverName);
        }
        catch (ClassNotFoundException ex){
            System.out.println("mysql 驱动找不到!");
        }
    }
    /*和指定数据库建立连接*/
    public Connection getConnection() {
      try{
        con=DriverManager.getConnection(dbUrl, dbUser, dbPwd);
      }catch (SQLException e){}
      return con;
    }
    /*关闭连接*/
    public void closeConnection(Connection con) throws SQLException{
        con.close();
    }
}
```

(2) 用户登录的数据库访问类 UserLoginDAO。

```java
import java.sql.*;
public class UserLoginDAO {
    /*根据用户名和密码判断用户是否登录成功*/
    public int getLoginState(String loginName,String password){
    try{
        LinkDB db=new LinkDB();
        Connection conn=db.getConnection();    //连接数据库
        Statement st=conn.createStatement();
        String sqlStr="select * from users where loginname='"+loginName+"'";
        ResultSet rs=st.executeQuery(sqlStr);
        if(!rs.first()){
          return -1;                            //用户名不存在,返回-1
        }
        else{
          if(rs.getString("password").equals(password)){
            db.closeConnection(conn);           //关闭数据库连接
            return 1;                           //用户名和密码都匹配,返回1,说明登录成功
```

```
            }
        else{
            return 0;                          //密码不正确,返回 0
        }
      }
    }catch(SQLException esql){ }
        return 0;
    }
}
```

(3)用户登录界面的类 UserLogin。

```
import java.awt.*;
import java.awt.event.ActionEvent;
import java.awt.event.ActionListener;
import java.awt.event.ItemEvent;
import java.awt.event.ItemListener;
import javax.swing.*;
public class UserLogin extends JFrame implements ActionListener,ItemListener{
    private JTextField tf;                     //定义用户名输入框变量
    private JComboBox combox;                  //定义用户类型变量
    private JPasswordField pf;                 //定义密码输入框变量
    private JButton[] jb;                      //定义三个按钮的数组
    private JOptionPane op;                    //定义弹出对话框变量
    private int userType;                      //用户类型,0 为教师,1 为学生
    public UserLogin(){
        super("用户登录");
        tf=new JTextField();
        combox=new JComboBox();
        pf=new JPasswordField();
        jb=new JButton[3];
        //这里要注意一定要把对象数组的成员实例化
        jb[0]=new JButton("登录");
        jb[1]=new JButton("取消");
        jb[2]=new JButton("退出");
        op=new  JOptionPane();
        Container c=getContentPane();
        c.setLayout(new BorderLayout());
        JPanel [] jp={new JPanel(),new JPanel()};
        String []s={"Center","South"};
        for(int i=0;i<s.length;i++){
           c.add(jp[i],s[i]);
        }
        jp[0].setLayout(new GridLayout(3,2));     //设置网格布局方式
        jp[0].add(new JLabel("用户类型   ",SwingConstants.RIGHT));
        combox.addItem("教师用户");
```

```java
    combox.addItem("学生用户");
    combox.addItemListener(this);
    jp[0].add(combox);
    jp[0].add(new JLabel("用户名   ",SwingConstants.RIGHT));
    jp[0].add(tf);
    jp[0].add(new JLabel("密    码   ",SwingConstants.RIGHT));
    jp[0].add(pf);
    /*在jp[1]中将加入三个按钮,并注册事件监听*/
    jb[0].addActionListener(this);
    jb[1].addActionListener(this);
    jb[2].addActionListener(this);
    for(JButton b:jB.{
       jp[1].add(B.;
    }
    setBounds(200,200,220,180);            //设置登录界面显示位置(200,200)和大小
    setVisible(true);
}
/*获取选择用户类型的索引值*/
public void itemStateChanged(ItemEvent e){
    userType=combox.getSelectedIndex();
}
/*三个按钮的业务处理*/
public void actionPerformed(ActionEvent e){
    if(e.getSource()==jb[0]){              //登录按钮的处理
      if(tf.getText().equals("")){
         op.showMessageDialog(null,"用户名不可为空");
         return;
      }
      UserLoginDAO user=new UserLoginDAO();
      int isLogin=user.getLoginState(tf.getText().trim(), pf.getText().trim());
      switch(isLogin){
        case 1:
          if(userType==0)                  //若i==0,则是教师用户
             op.showMessageDialog(null,"教师用户登录成功");
          else
             op.showMessageDialog(null,"学生用户登录成功");
          break;
        case 0:
             op.showMessageDialog(null,"用户名或者密码不正确");
          break;
        default:
             op.showMessageDialog(null,"用户名不存在");
          break;
      }
    }
```

```java
            if(e.getSource()==jb[1]){                //取消按钮的处理
                tf.setText("");
                pf.setText("");
                return;
            }
            if(e.getSource()==jb[2]){                //退出按钮的处理
                System.exit(0);
            }
        }
        public static void main(String args[]){
            UserLogin login=new UserLogin();
            login.setDefaultCloseOperation(JFrame.EXIT_ON_CLOSE);
        }
    }
```

12.3 练 习 题

12.3.1 选择题

1. 关于 JDBC 访问数据库的说法错误的是( )。
   A. 建立数据库连接时,必须加载驱动程序,可采用 Class.forName()实现
   B. 建立与某个数据源的连接可采用 DriverManager 类的 getConnection 方法
   C. 建立数据库连接时,必须要进行异常处理
   D. JDBC 中查询语句的执行方法必须采用 Statement 类实现
2. JDBC 驱动的种类有( )种。
   A. 2   B. 3   C. 4   D. 5
3. 下列哪些是 JDBC 驱动程序?( )
   A. JDBC-ODBC 桥和 ODBC 驱动程序
   B. 本地 API 部分用 Java 来编写的驱动程序
   C. JDBC 网络纯 Java 驱动程序
   D. 本地协议纯 Java 驱动程序
4. 下面哪个方法可以用来加载 JDBC 驱动程序?( )
   A. 类 java.sql.DriverMananger 的方法 getDriver
   B. 类 java.sql.DriverMananger 的方法 getDrivers
   C. 类 java.sql.Driver 的方法 connect
   D. 类 java.lang.Class 的方法 forName
5. 下面哪些方法可以用来建立数据库连接?( )
   A. 类 java.sql.DriverMananger 的方法 getConnection
   B. 类 javax.sql.DataSource 的方法 getConnection
   C. 类 javax.sql.DataSource 的方法 connection
   D. 类 java.sql.Driver 的方法 getConnection

6. 关于 Statement 常用的方法,下列说法哪个是不正确的?(　　)

   A. 方法 execute 和 executeQuery 一样都返回单个结果集

   B. 方法 executeUpdate 用于执行 insert、update 或 delete 语句以及 SQL DDL(数据定义语言)语句

   C. 方法 execute 用于执行返回多个结果集、多个更新计数或两者组合的语句

   D. 方法 executeQuery 用于产生单个结果集的语句

7. 关于 Statement 的使用,下列说法正确的是(　　)。

   A. 继承了 Statement 接口中所有方法的 PreparedStatement 接口都有自己的 executeQuery、executeUpdate 和 execute 方法

   B. Statement 是预编译的,效率高

   C. Statement 对象用类 Connection 的方法 createStatement 创建

   D. 还可以绑定参数,防止 SQL 注入问题

8. 关于 PreparedStatement 对象的使用,下列说法不正确的是(　　)。

   A. PreparedStatement 是个类

   B. PreparedStatement 继承了 Statement

   C. PreparedStatement 是预编译的,效率高

   D. PreparedStatement 可以绑定参数,防止 SQL 注入问题

9. 下列关于 ResultSet 接口的说法正确的是(　　)。

   A. ResultSet 接口被用来提供访问查询结果的数据表,查询结果被当做 ResultSet 对象而返回

   B. ResultSet 对象提供"指针",指针每次访问数据库表的一行

   C. ResultSet 的 next()方法用来移动指针到数据表的下一行,如果到达表尾,next()方法返回 false,否则为 true

   D. ResultSet 接口提供大量的获得数据的方法,这些方法返回数据表中任意位置的数据,不论是基本数据类型或引用数据类型的数据

10. 能够完成更新的 ResultSet 的 SQL 语句必须要具备(　　)。

    A. 只引用了单个表

    B. 可以是多个表

    C. 不含有 join 或者 group by 子句

    D. 列中要包含主键

11. 以下关于 CallableStatement 对象说法正确的是(　　)。

    A. CallableStatement 对象用于执行数据库中存储过程的调用

    B. CallableStatement 对象中,有一个通用的成员方法 call,这个方法用于以名称的方式调用数据库中的存储过程

    C. 在数据库调用过程中,可以通过设置 IN 参数向调用的存储过程提供执行所需的参数

    D. 在存储过程的调用中,通过 OUT 参数获取存储过程的执行结果

E. CallableStatement 对象用于执行对数据库所有的调用

12. 方法 executeUpdate 能用于的 SQL 语句有（   ）。
    A. insert　　　　　B. select　　　　　C. update　　　　　D. delete
13. 调用数据库的存储过程或函数用以下哪个类？（   ）
    A. CallableStatement　　　　　　　B. Statement
    C. PreparedStatement　　　　　　　D. Connection
14. 以下哪个方法不是定义在 ResultSet 中用于 Cursor 定位的方法？（   ）
    A. next()　　　　　　　　　　　　B. beforeFirst()
    C. afterLast()　　　　　　　　　　D. isBeforeFirst()
15. 关于 JDBC API 管理的事务下列说法不正确的是（   ）。
    A. 默认情况下，新连接将处于自动提交模式
    B. 方法 commit 使 SQL 语句对数据库所做的任何更改成为永久性的，它还将释放事务持有的全部锁，而方法 rollback 将弃去那些更改
    C. 默认情况下，新连接将处于手动提交模式
    D. 如果两个更新都是成功，则调用 commit 方法，从而使两个更新结果成为永久性的；如果其中之一或两个更新都失败了，则调用 rollback 方法，以将值恢复为进行更新之前的值

12.3.2 编程题

1. 已知某学生管理信息系统数据库 stuInfoManage 里面有一张学生表 student，student 的各字段含义如表 12-3 所示。

表 12-3　student 的各字段

字段名称	说明	字段类型	是否为空	备注
stuID(PK)	学号	Varchar(50)	否	
name	学生姓名	Varchar(50)	否	
sex	学生性别	Varchar(10)	否	
birthDate	出生日期	Varchar(50)	否	格式为 1986-1-1
class	班级	Varchar(50)	否	
zhuanye	专业	Varchar(50)	否	
grade	入学的年级	int	否	如 2004 级

请编写一个 Java 程序 AddStudentInfo.java 将学生名字为"张三"的信息插入表中，信息如表 12-4 所示。

表 12-4　信息表

学号	姓名	性别	出生日期	班级	专业	入学年份
20122660801	张三	男	1994-10-9	软工 12(8)	软件工程	2012

2. 编写一个从数据库的学生表中查询 20 岁以下的学生信息的程序 SearchStudentInfo.java 并显示查询结果,其中数据库名为 stuInfoManage,学生表名为 student,student 的各字段含义如表 12-5 所示。

表 12-5 student 各字段的含义

字段名称	说明	字段类型	是否为空	备注
stuID(PK)	学生学号	Varchar(50)	否	
name	学生姓名	Varchar(50)	否	
sex	学生性别	Varchar(10)	否	
birthDate	出生日期	Varchar(50)	否	格式为 1986-1-1
class	班级	Varchar(50)	否	
zhuanye	专业	Varchar(50)	否	
grade	入学的年级	int	否	如 2012 级

3*. 编写一个简单的通讯簿,要求可对朋友的姓名、性别、出生日期、工作单位、手机号码、联系电话、职称、职务、联系地址、邮编、E-mail 和 QQ 号等信息保存、查询、修改和删除等功能。系统功能界面图如图 12-2 所示。

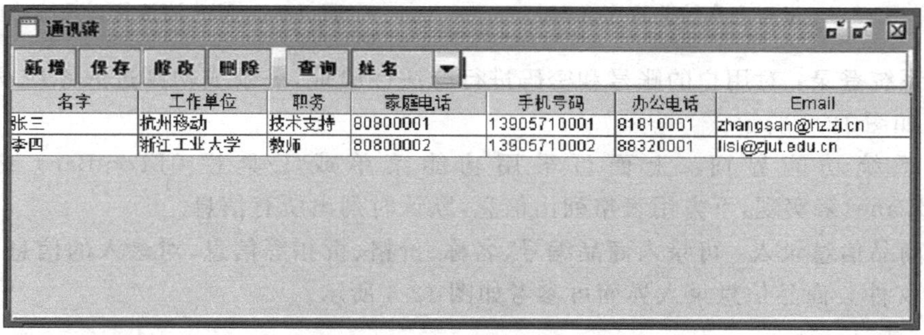

图 12-2 通讯簿界面

(1) 系统登录:对用户的账号和密码进行合法性验证,登录成功则进入系统功能界面。

(2) 系统功能界面:上面可采用功能菜单或工具栏(JToolBar)或选项板(JTabbedPane)来实现,下方用表格列出信息,默认时列出所有信息。

(3) 新增功能:可录入朋友的姓名、性别、出生日期、工作单位、手机号码、联系电话、职称、职务、联系地址、邮编、E-mail 和 QQ 号等信息。

(4) 保存功能:对录入的信息保存至数据库表。

(5) 修改功能:可对朋友的姓名、性别、出生日期、工作单位、手机号码、联系电话、职称、职务、联系地址、邮编、E-mail 和 QQ 号等信息进行修改。

(6) 删除功能:可选中表格中的某个信息进行删除。

(7) 查询功能:可根据姓名、性别、出生日期、工作单位、手机号码、联系电话、职称、职务、联系地址、邮编、E-mail 和 QQ 号等条件进行查询朋友的信息。

4*. 目前与观众互动类电视节目越来越多,如一场球赛转播,观众可以发短信与主持人进行评球,转播结束后,主持人利用抽奖程序,从发送短信的观众的手机号码中抽取若干个号码作为幸运观众,并给予一定的奖品。请编写一个简单的幸运观众抽奖程序,要求设定简单的参数后,能随机抽取数据库中的若干个观众手机号码,显示时隐藏最后两位号码,并同时显示该手机号码的所属地,界面如图12-3所示。

5*. 编写一个简单的商场 VIP 消费管理系统,系统功能包括商品信息录入、商品信息查询、VIP 信息录入、VIP 消费记录查询、VIP 消费购物记录登记、系统维护(系统用户、商品和 VIP 信息等实现管理)和系统帮助。系统功能框架图如图12-4所示。

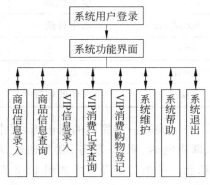

图 12-3　幸运观众抽奖程序界面　　　　图 12-4　VIP 消费管理系统功能框图

（1）系统登录:对用户的账号和密码进行合法性验证,登录成功则进入系统功能界面。登录界面如图12-5所示。

（2）系统功能界面:上面可采用功能菜单或工具栏(JToolBar)或选项板(JTabbedPane)来实现,下方用表格列出信息,默认时列出所有信息。

（3）商品信息录入:可录入商品编号、名称、价格、折扣等信息,对录入的信息保存至数据库表或文件。商品信息录入界面可参考如图12-6所示。

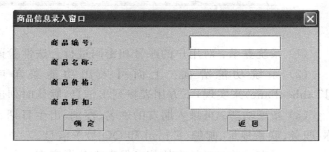

图 12-5　登录界面　　　　　　　　　图 12-6　商品信息录入界面

（4）商品信息查询:可根据商品编号或名称进行查询。商品信息查询界面可参考图12-7。

（5）VIP 信息录入:可对 VIP 用户的姓名、年龄、入会时间、性别、工作单位、手机号码、联系电话、联系地址、邮编等信息进行输入。

（6）VIP 消费记录查询:可对 VIP 用户的消费记录进行查询。VIP 消费记录查询界面

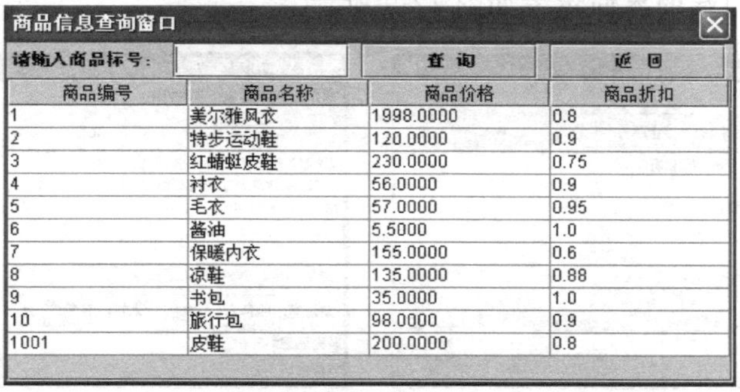

商品信息查询窗口

请输入商品标号: 查询 返回

商品编号	商品名称	商品价格	商品折扣
1	美尔雅风衣	1998.0000	0.8
2	特步运动鞋	120.0000	0.9
3	红蜻蜓皮鞋	230.0000	0.75
4	衬衣	56.0000	0.9
5	毛衣	57.0000	0.95
6	酱油	5.5000	1.0
7	保暖内衣	155.0000	0.6
8	凉鞋	135.0000	0.88
9	书包	35.0000	1.0
10	旅行包	98.0000	0.9
1001	皮鞋	200.0000	0.8

图 12-7　商品信息查询界面

如图 12-8 所示。

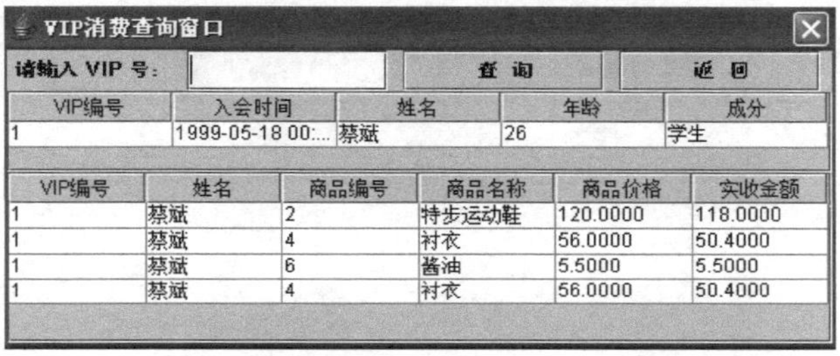

VIP消费查询窗口

请输入 VIP 号: 查询 返回

VIP编号	入会时间	姓名	年龄	成分
1	1999-05-18 00:...	蔡斌	26	学生

VIP编号	姓名	商品编号	商品名称	商品价格	实收金额
1	蔡斌	2	特步运动鞋	120.0000	118.0000
1	蔡斌	4	衬衣	56.0000	50.4000
1	蔡斌	6	酱油	5.5000	5.5000
1	蔡斌	4	衬衣	56.0000	50.4000

图 12-8　VIP 消费记录查询

(7) VIP 消费购物记录登记：可对用户的购物记录进行登记，保存至数据库或文件。

(8) 系统维护包括对系统用户、商品和 VIP 信息等实现管理。

(9) 系统帮助：提供系统的帮助信息。

6*. 利用基于 TCP 的 Socket 编程、多线程和 JDBC 实现一个简单的聊天系统，包括服务器端程序和客户端程序。

(1) 服务器端程序要求如下。

① 能等待用户联机，并建立和客户端通信的 I/O 通道。

② 当有用户要求加入聊天室，先验证其账号和密码的合法性，验证成功后允许联机并将登录信息写入日志 log 表。

③ 能将用户发来的信息存入数据库保存，以便查询统计，同时能将消息广播给所有聊天室成员。

④ 能清除不正常终端的联机。

⑤ 系统管理：能设置加入聊天系统的人数限制，能查看用户登录信息和聊天信息。

(2) 客户端程序要求如下。

① 能注册和登录账号。

② 能与服务器程序联机并且建立输入输出通道，通道建立完成后，其工作便是接收由客户端传来的消息，然后根据消息来作出适当的处理。

③ 列出在线的所有用户登录名或昵称。

服务器端和客户端的界面参考如图 12-9 所示。

图 12-9　简单的聊天系统界面

第 13 章　XML 及程序打包

13.1　知　识　点

（1）XML 在 Java 程序中的应用。
（2）DOM 编程。
（3）Java 程序的发布,包括利用 cmd 工具打包和利用 Eclipse 打包。
重点：XML 在 Java 程序中的应用,DOM 编程,利用 Eclipse 打包。
难点：DOM 编程。

13.2　例　　题

【例 13-1】　简述 Java 解析 XML 的四种方法：DOM、SAX、JDOM 和 DOM4J。
【解析】　Java 解析 XML 的四种常见方法包括 DOM、SAX、JDOM 和 DOM4J。

（1）DOM 方法采用拉模型,把整个文档加载到内存中,便于操作。它支持删除、修改、重新排列等多种功能。缺点是将整个文档调入内存会浪费时间和空间；适用于一旦解析了文档还需多次访问这些数据且硬件资源(内存、CPU)充足的环境。

（2）SAX 方法采用推模型、事件驱动编程。当解析器发现元素开始、元素结束、文本和文档的开始或结束等时,发送事件,程序员编写响应这些事件的代码,保存数据。它的优点是不用事先调入整个文档,占用资源少；缺点是数据保存不持久,从事件中只能得到文本,但不知该文本属于哪个元素；适用于的场合包括数据量较大的 XML 文档,占用内存高,机器内存少,无法一次加载 XML 到内存；只需 XML 文档的少量内容,很少重复访问。

（3）JDOM 的目的是成为 Java 特定文档模型,它简化与 XML 的交互并且比使用 DOM 实现更快。JDOM 自身不包含解析器,它通常使用 SAX2 解析器来解析和验证输入 XML 文档。它的优点是极大减少了代码量,提供常用 API 减少重复劳动。

（4）DOM4J 是一种解析 XML 文档的开放源代码 XML 框架,具有性能优异、功能强大和极易使用的特点。DOM4J 最大的特色是使用大量的接口,这也是它被认为比 JDOM 灵活的主要原因。

【例 13-2】　以下哪个是 Java 程序发布的命令？
A. package　　　　　B. jar　　　　　　C. javadoc　　　　　D. java
【解析】　Java 程序发布可用 jar 命令。本题选 B。

13.3　练　习　题

13.3.1　判断题

1. DOM 解析是将 XML 文件全部载入,组装成一颗 DOM 树,然后通过节点以及节点之间的关系来解析 XML 文件。   (   )
   2. DOM 和 SAX 解析都不适用于大数据量的 XML 文件。   (   )
   3. DOM 解析 XML 时,需加装 JDK 自带的 org.w3c.dom、org.xml.sax 和 javax.xml.parsers 包。   (   )
   4. XML 是一种数据表示的格式,利用 XML 可以方便地对数据进行组织和存储。
                                                                        (   )
   5. 当一个 Java 项目完成后,可采用 jar 命令、Eclipse 编程工具、专用打包工具打包等进行打包发布。   (   )

13.3.2 选择题

1. XML 是(   )。
   A. 一种标准泛用标记语言   B. 一种扩展性标识语言
   C. 一种超文本标记语言   D. 一种层叠式表单标识语言
2. 关于 XML 的用途有哪些?(   )
   A. XML 能够存储 HTML 显示的文件内容
   B. XML 可以作为数据交换的格式
   C. XML 可以作为数据存储的格式
   D. XML 被广泛应用在电子商务、电子政务的应用系统中
3. 如果需要在 XML 文件中显示中文,那么设置 encoding 的值为(   )。
   A. GB2312   B. UTF-8   C. BIG5   D. UTF-16
4. <Name StudentID="2012001">Tom</Name>中,以下数据部分是(   )。
   A. Name   B. StudentID   C. 2012001   D. Tom
5. 已知某 XML 文档中声明<? xml version="1.0">,则该文档采用什么编码标准?
(   )
   A. GB2312   B. UTF-16   C. UTF-8   D. ASCII
6. 在 XML 文档中包含多个重数值的是(   )。
   A. 属性   B. 子元素   C. 命名空间   D. 标记
7. 采用 DOM 解析 XML 文件一般需加载哪些包?(   )
   A. org.w3c.dom   B. org.xml.sax
   C. javax.xml.parsers   D. javax.xml
8. 在 cmd 命令窗口下可使用以下哪个命令可以将已编译的 java 类打包?(   )
   A. jar   B. java   C. javac   D. package

第二部分

参考答案

本部分给出所有练习题的参考答案,部分容易出错的题目给出了解析。因篇幅有限,一些编程题参考代码不附在书上。书中的程序源代码和编程题参考代码可从本书配套网站(http://www.zjut-java.com)或清华大学出版社网站(http://www.tup.com.cn)下载。

第三部分

参考答案

第 14 章　Java 语言概述参考答案

14.1 判　断　题

1. 错　2. 对　3. 对　4. 对　5. 错　6. 错　7. 对　8. 对　9. 对　10. 对

14.2 选　择　题

1. D　2. B　3. B　4. C　5. C　6. C　7. C　8. A　9. BC
10. B　11. B　12. B　13. A　14. B　15. A　16. D　17. BC

【解析】

第 8 题：本题考查 Java 运行的机制，属于难点。类加载器为程序的执行加载所需要的全部类，它将本地文件系统的类名和从网络导入的类名空间相分离，本地类总是首选被加载以增加安全性。字节代码校验器只要负责基于代码的规范，包括语法语义的检查和安全性检查。Java 运行时解释器是 JVM 的核心，实现把抽象的字节码指令映射到本地系统平台下的库引用或指令。本题选 A。

第 11 和 12 题：这 2 题考查类名和文件名的命名规范，是 Java 初学者容易忽视的问题。在 Java 中，一个 Java 源程序可以包含多个 class 类，但只能有一个 public 修饰的类定义，此时保存源程序的文件名与该类名相同，扩展名必须是 java。

若一个 Java 源程序文件中只有一个类（如 Demo），则源文件名与该类名一致，保存为 Demo.java；若一个 Java 源文件有 2 个或 2 个以上类组成，则有以下三种情况。

(1) 若有一个类为 public 修改，则源文件名必须与该类名一致。

(2) 若有一个含有 main 方法的类名，则源文件名必须与该类名一致。

(3) 若没有一个类是 public 类，而且没有一个类含有 main 方法，则可取该源文件中任意一个类名作为源文件名。

第 14 题：本题考查 Java 应用程序运行的入口方法。Java 应用程序必须要有 main 方法才允许运行，其中 main 方法必须声明为 public static void main(String args[])，这里唯一能修改的是参数数组名（比如 args 改为 a）和字符串数组的写法（比如可写成 String [] args)，否则就无法执行。本题的 main 方法在 void 前没有 static 修饰，编译不会出错，但程序不能正常运行，会出现异常。

14.3 简　答　题

1. 简述 Java 程序的可移植性。

答：与平台无关的特性使 Java 程序可以方便地被移植到网络上的不同机器。同时，Java 的类库中也实现了与不同平台的接口，使这些类库可以移植。另外，Java 编译器是由

Java 语言实现的,Java 运行时系统由标准 C 实现,这使得 Java 系统本身也具有可移植性。

2. Java 程序是由什么组成的？Java 源文件的命名规则是怎样的？

答：一个 Java 程序是由若干个类组成的,但只能有一个类是 public 类。Java 源文件命名规则是：源文件名必须与其中的 public 类的名字相同,扩展名是 java；如果源文件中没有 public 类,那么源文件的名字只要和某个类的名字相同,并且扩展名是 java 就可以了。

第 15 章　Java 基础语法参考答案

15.1　判　断　题

1. 对　2. 对　3. 错　4. 对　5. 错

【解析】

第 3 题：本题考查 Java 中的流程控制语句。Java 提供了循环控制（while、do-while 和 for）、选择控制（if 和 switch）、转向控制（break、continue 和 return）和异常处理等几种流程控制语句。

第 5 题：本题考查算术运算中的数据类型转换。在算术运算中，若一个操作数是整型，另一个操作数是浮点类型，则结果也是浮点类型。本题中的 int 应改为 float。

15.2　选　择　题

1. D	2. B	3. C	4. D	5. D	6. B	7. B	8. A	9. D
10. B	11. C	12. C	13. D	14. D	15. D	16. D	17. D	18. A
19. B	20. B	21. B	22. B	23. D	24. D	25. D	26. C	27. C
28. B	29. A	30. C	31. B	32. B	33. A	34. C	35. A	

【解析】

第 9 题：在 Java 中，char 类型的值可转化为 ACSII 值，所以 char 类型和 int 类型可分别作为操作数执行＋、－、×、/、%等算术运算，运算的结果若赋值给 char 类型变量，则结果也是 char 类型，否则是 int 类型。本题选项 D，字符'1'的 ACSII 值是 49，运算后 k 的值为 50。选项 B 应该声明为"float d＝45.6f;"。

15.3　程序阅读题

1. (1) 输出 10 以内的素数
   (2) 3 5 7
2. i＝1 j＝3
   i＝1 j＝2
   i＝2 j＝3

15.4　编　程　题

1. 参考代码如下：

```
public class PrintStuScore {
```

```java
    public static String scoreLevel(int score) {
        if(score>=90){
            return "优";
        }
        else if(score>=80){
            return "良";
        }
        else if(score>=70){
            return "中";
        }
        else if(score>=60){
            return "及格";
        }
        else{
            return "不及格";
        }
    }
    public static void main (String[] args) {
        int javaScore=90;
        int databaseScore=75;
        int englishScore=85;
        int avrScore= (javaScore+databaseScore+englishScore)/3;
        System.out.println("Java课程成绩："+javaScore+" "+scoreLevel(javaScore));
        System.out.println("数据库课程成绩："+databaseScore+" "+scoreLevel
            (databaseScore));
        System.out.println("英语课程成绩："+englishScore+" "+scoreLevel(englishScore));
        System.out.println("平 均 成 绩："+avrScore);
    }
}
```

2. 参考代码如下：

```java
public class FindPrime{
    public static void main(String[] args) {
        int primeNumber=0;                    //素数个数
        next:for( int i=101;i<=200;i++){
            for (int j=2;j<i;j++){
                if (i%j==0){          //如果j是i的因子,则i不是素数,取下一个数进行判断
                    continue next;
                }
            }
            primeNumber++;
            System.out.print(i+" ");          //输出素数
            if(primeNumber%10==0){
                System.out.println();
            }
```

            }
        }
    }

3.（方法一）采用 if 语句实现，参考代码如下：

```java
public class ExeIF {
    public static void main(String args[]){
        double t,y;
        t=2.5;
        //t=Double.parseDouble(args[0]);从命令行参数读入 t 的值
        y=0.0;
        if(t>=0&&t<1){
            y=Math.pow(t,2.0)-1;
        }
        else if(t>=1&&t<3){
            y=Math.pow(t,3.0)-2*t-2;
        }
        else if(t>=3&&t<5){
            y=Math.pow(t,2.0)-t*Math.sin(t);
        }
        else if(t>=5&&t<7){
            y=++t;
        }
        else{
            y=--t;
        }
        System.out.println("y="+y);
    }
}
```

（方法二）采用 switch 语句实现，参考代码如下：

```java
public class ExeSwitch {
    public static void main(String args[]){
        double t=2.5,y=0.0;
        switch((int)t){
            case 0:y=Math.pow(t,2.0)-1; break;
            case 1:
            case 2:y=Math.pow(t,3.0)-2*t-2;break;
            case 3:
            case 4:y=Math.pow(t,2.0)-t*Math.sin(t);break;
            case 5:
            case 6:y=++t;break;
            default: y=--t;
        }
        System.out.println("y="+y);
```

        }
    }

4. 参考代码如下：

```
class ForDemo{
    public static void main(String args[]){
        int n=0;
        for(int i=10;i<=100;i++){
            if(i%2==0&&i%3!=0){
                n++;
                System.out.print(i+" ");
            }
            if(n>=10){
                n=0;
                System.out.println();
            }
        }
    }
}
```

5. 参考代码如下：

```
import java.util.*;                              //调用随机数类 Random
public class RandomSum {
    public static void main(String args[]){
        Random r=new Random();
        int n=r.nextInt(10);                     //随机产生一个 0~10 的随机数
        int sum=0;
        for(int i=0;i<n;i++){                    //循环产生 n 个随机数
            int a=r.nextInt(100);                //生成一个 0~100 的整数
            System.out.print(a+" ");
            sum+=a;
        }
        System.out.print("\n随机数和为："+sum);
    }
}
```

6. 参考代码如下：

```
public class SisterPrime{
    public static void main(String args[]){
        int number[]=new int[900];               //用来存放 100~1000 的整数
        int a=100;
        int count=0;                             //计数
        for(int i=0;i<900;i++){
            number[i]=a++;
        }
```

```java
            for(int i=0;i<900;i++){
                if(prime(number[i])&&prime(number[i+2])){
                    count++;
                    System.out.print(number[i]+" "+number[i+2]+" ");
                    if(count%10==0)                  //每行输出10对姐妹素数
                        System.out.println();
                }
            }
        }
    }
    public static boolean prime(int num){
        boolean flag=true;                           //判断是否为素数的变量
        for (int i=2;i<=(num-1);i++){                //从2开始至当前整数-1循环
            if (num%i==0) {
                //如果从2开始循环到有整数能被当前整数整除,则当前数不为素数,跳出循环
                flag=false;                          //设置flag为false
                break;
            }
        }
        return flag;
    }
}
```

7. 参考代码如下：

```java
public class DaffodilNumber {
    public static void main(String args[]){
        System.out.println("水仙花数：");
        int number[]=new int[900];
        int i=100;
        for(int a:number){                           //for语句简化写法
            a=i++;
            int b11=a/100;                           //百位数
            int b12=a%100;
            int b21=b12/10;                          //十位数
            int b22=b12%10;                          //个位数
            if(a==(Math.pow(b11, 3)+Math.pow(b21, 3)+Math.pow(b22, 3)))
                System.out.print(a+" ");
        }
    }
}
```

8. 参考代码如下：

```java
public class CompleteNumber {
    public static void main(String args[])    {
        System.out.print("1000之内所有完全数:"+"\n"+1+" ");
        for(int i=2;i<1000;i++){
```

```
            int sum=0;
            for(int j=1;j<i;j++){
                if(i%j==0)
                    sum+=j;
            }
            if(i==sum)
                System.out.print(i+" ");
        }
    }
}
```

9. 参考代码如下：

```
public class CalculateE {
    public static void main(String args[]){
        double sum=0;
        int n=1;
        for(int i=1;i<=n;i++){
            sum+=1/product(i);
            n++;
            if(1/product(n)<0.0001)          //当 1/n<0.0001 跳出循环即停止相加
                break;
        }
        double e=sum+1;
        System.out.print("e="+e);
    }
    public static double product(int y) {    //计算 y 的阶乘
        double a=1.0;
        for(int i=y;i>=1;i--)   {
            a*=i;
        }
        return a;
    }
}
```

10. 参考代码如下：

```
import java.util.Scanner;
public class Cos {
    public static void main(String args[]){
        Scanner sc=new Scanner(System.in);              //从键盘获取 x 的值
        double x=sc.nextDouble();
        double sum=0;
        for(int i=1;i<7;i++){                           //这里要注意循环终止条件
            sum+=Math.pow(-1,i) * Math.pow(x,2*i)/product(2*i);
        }
```

```
        System.out.print((double)(Math.round((sum+1) * 1000000))/1000000);
                                                //精确到 0.000001
    }
    public static double product(int y) {       //计算 y 的阶乘
        double a=1.0;
        for(int i=y;i>=1;i--){
            a*=i;
        }
        return a;
    }
}
```

第 16 章　类和对象参考答案

16.1　判　断　题

1. 对　2. 错　3. 对　4. 错　5. 对　6. 对　7. 对　8. 对　9. 对　10. 对
11. 对　12. 对　13. 对　14. 错　15. 对

【解析】

第 2 题：本题考查构造方法的特性。如果定义的类中没有给出构造方法，系统会加载一个默认的构造方法，默认的构造方法没有参数，方法体中也没有语句。

第 4 和 5 题：考查关键字 static 修饰的方法。非 static 修饰的方法称为对象方法，必须实例化为对象后，才能被对象调用，不能用类名直接调用。

16.2　选　择　题

1. A　2. C　3. B　4. A　5. D　6. A　7. D　8. D　9. A
10. C　11. B　12. B　13. BC　14. C　15. C　16. B　17. A　18. D
19. C　20. A　21. C　22. C　23. B　24. B　25. C

【解析】

第 6 题：本题考查 Java 的垃圾内存回收机制。本题 main 方法中用"new Book(3);"创建的实例对象，由于没有引用，该对象占用的存储单元属于垃圾内存，在申请立即回收垃圾 System.gc()语句执行时，会自动调用 finalize 方法，此时 id 值为 3。所以，本题选 A。

第 7 题：本题考查关键字 final 的用法。final 可以修饰类、成员方法、成员变量以及方法中的参数等，但不能修饰接口。final 修饰的类不能被继承，final 修饰的方法不能被子类覆盖，final 修饰的变量是常量，不能修改其值。

第 13、24 和 25 题：这几题考查 static 的用法。static 修饰的成员变量和成员方法分别称为类变量和类方法，没有 static 修饰的成员变量和成员方法称为实例变量和实例方法。类方法中可直接调用类变量和类方法，不能直接调用实例变量和实例方法，反之，实例方法中可直接调用实例变量和实例方法，不能直接调用类变量和类方法。

第 23 题：本题考查 package 的用法。一个 Java 源程序文件中，只能有一个 package 语句，该语句只能放在程序除注释以外的第一行。本题选 B。

16.3　程序阅读题

1. 第 6 行出现编译出错，理由是第 6 行的 a+b 结果为 int 类型，而类中没有定义 int 类型参数的构造方法。

2. 第 4 行出现编译出错，理由是第 9 行定义的 fly 方法修饰为 private，无法在另一个类中访问。

3. x=2
   st.y=1
   x=2
   st.y=1
4. Int value is:33
   Str value is:Hi
   Pt value is:22.0
5. ss=浙江工业大学 ms=软件学院
   ss=浙江工业大学 ms=计算机学院
6. (1) 两个类不在同一个源程序文件。
   (2) 编译正常,运行结果为 I am ClassA.
7. 第10行的方法改为 public String toString()
   第16~19行语句改为 Person p=new Person("张三",20,"男");
   第20行语句改为 System.out.println(p.toString());
8. 7

16.4 编 程 题

1. 参考代码如下：

```
public class Suansu {
  int a,b;
  public Suansu(){
    a=10;
    b=5;
  }
  public int addAB(){
    return a+b;
  }
  public int subAB(){
    return a>b?a-b:b-a;
  }
  public int multiAB(){
    return a*b;
  }
  public double Divab(){
    return (double)a>b?a/b:b/a;
  }
}
class TestSuansu{
  public static void main(String args[]){
```

```java
        Suansu s=new Suansu();
        System.out.println(s.addAB());
        System.out.println(s.subAB());
        System.out.println(s.multiAB());
        System.out.println(s.Divab());
    }
}
```

2. 参考代码如下：

```java
class Rectangle{
    private int width, height;
    public Rectangle(int w,int h){
        width=w;
        height=h;
    }
    public Rectangle(){
        width=4;
        height=5;
    }
    public int getWidth(){
        return width;
    }
    public int getHeight(){
        return height;
    }
    public void setWidth(int w){
        if (w>0 && w<30){
            width=w;
        }
        else{
            System.out.println("Invalid Width");
        }
    }
    public void setHeight(int h){
        if (h>0 && h<30){
            height=h;
        }
        else{
            System.out.println("Invalid height");
        }
    }
    public int getPerimeter(){
        return width +width +height +height;
    }
    public int getArea(){
```

```java
            return width * height;
        }
        public void draw(){                    //画矩形
            for (int rowCounter=0; rowCounter<height; rowCounter++){
                for (int colCounter=0; colCounter<width; colCounter++)
                    if(rowCounter==0||(rowCounter==height-1)||colCounter==0||(colCounter==width-1)){
                        System.out.print("*");
                    }
                    else{
                        System.out.print(" ");
                    }
                System.out.println();
            }//end for
        }//end draw
    }//end class
public class TestRectangle {
    public static void main(String args[]){
        Rectangle r=new Rectangle(5,6);
        System.out.println("矩形的面积是:"+r.getArea());
        System.out.println("该矩形的形状如下:");
        r.draw();
    }
}
```

3. 参考代码如下:

```java
public class Student {
    private String name;
    private int age;
    public Student(){
        name="无名氏";
        age=20;
    }
    public void setName(String name){
        this.name=name;
    }
    public String getName(){
        return name;
    }
    public void setAge(int age){
        this.age=age;
    }
    public int getAge(){
        return age;
    }
```

```java
        public boolean isSameAge(Student s){
            if(this.age==s.age)
                return true;
            return false;
        }
        public static void main(String args[]){
            Student s1=new Student();
            s1.setName("张三");
            s1.setAge(18);
            System.out.println("name: "+s1.getName());
            System.out.println("age: "+s1.getAge());
            Student s2=new Student();
            s2.setAge(19);
            System.out.println("Same age?"+s1.isSameAge(s2));
        }
    }
```

4. 略

5. 参考代码如下：

```java
public class MyPoint{
    private int x,y;
    public MyPoint(){
        x=0;
        y=0;
    }
    public MyPoint(int x,int y){
        this.x=x;
        this.y=y;
    }
    public float getD(MyPoint p){
        return (float)Math.sqrt((p.x-x) * (p.x-x)+(p.y-y) * (p.y-y));
    }
    public static void main(String args[]){
        MyPoint p1=new MyPoint(2,3);
        MyPoint p2=new MyPoint(4,5);
        System.out.println(p1.getD(p2));

    }
}
```

6. 略。

7. 参考代码如下：

(1) classA.java 的代码。

```java
package package1;
public class classA{
```

```
    public void methodA(){
        System.out.println("methodA");
    }
}
```

(2) classB.java 的代码。

```
package package2;
import package1.classA;
public class classB {
    public void methodB(){
        classA a=new classA();
        a.methodA();
        System.out.println("methodB");
    }
}
```

(3) 测试类 Test.java 代码。

```
package package2;
import package2.classB;
public class Test{
    public static void main(String args[]){
        classB b=new classB();
        b.methodB();
    }
}
```

8. 略。

第17章 类的封装性、继承性、多态性及接口参考答案

17.1 判 断 题

1. 对　2. 对　3. 错　4. 错　5. 对　6. 错　7. 对　8. 错
9. 对　10. 对　11. 错　12. 对　13. 错　14. 对　15. 对　16. 错
17. 对　18. 对　19. 错　20. 错

【解析】

第3题：本题考查抽象类的特性。抽象类中可包含构造方法、抽象方法和具体实现的方法以及常量和变量，但不能直接实例化，一般可对抽象类的子类实例化，实例化对象引用可以是抽象类类型。

第11题：本题考查子类中可以写的方法。子类可以继承父类的方法，也可以覆盖父类的方法，也可以新增加方法，新增加的方法与从父类继承的方法可构成重载。

第13题：本题考查abstract的用法。abstract可修饰类和方法，但不能修饰属性，即不能修饰类的成员变量。

第16题：本题考查final修饰的方法的特性。final修饰的方法不能被覆盖，但可以有重载的方法。

17.2 选 择 题

1. B　2. B　3. A　4. B　5. A　6. C　7. B　8. D
9. B　10. B　11. A　12. D　13. D　14. A　15. D　16. D
17. C　18. D　19. AD　20. CD　21. C　22. C　23. C　24. B
25. B　26. D　27. C　28. A　29. B　30. D

【解析】

第7题：本题考查对继承、覆盖和重载的理解，属于难点。本题4个选项中的方法名与父类定义的方法名都相同。选项A的方法有3个参数，选项C的方法有1个参数，都能与从父类继承的方法构成重载，可以加入子类Child中；选项B和选项D方法中的参数类型和个数都与父类定义的方法相同，由于父类定义的方法用public修饰，要想覆盖父类定义的方法，子类的方法也必须用public修饰，因此，选项D可以加入子类Child中，但选项B不行。

第8、10、13和14题：这几题主要考查子类中含有覆盖方法的调用情况。子类对象调用方法时，首先调用子类中同名和参数形式（参数个数、类型和顺序）一致的方法，否则就调用父类中定义的匹配的方法。

第9题：本题考查对类实现接口的理解。在Java的接口中包括抽象方法和常量。类实现接口时，需实现接口中的所有方法，并可直接使用接口中定义的常量（等同于从接口中继

承了这些常量)。本题中的接口 B 定义了"int k=10;"编译时会自动加上 final,表示 k 为常量。因此,本题选 B。

第 11 题:本题考查父类中有 static 修饰和没有 static 修饰的方法被子类覆盖的情形。当父类的方法被 static 修饰时,子类中覆盖的方法也必须用 static 修饰。反之,父类的方法没有 static 修饰,子类中覆盖的方法也不用 static 修饰。所以,本题选 A。

第 12 和 16 题:这 2 题考查子类继承父类时构造方法的定义。子类继承父类时,若父类定义了有参数的构造方法,而没有定义无参数的构造方法,那么在子类中的构造方法必须用 super 语句去调用父类定义的有参数的构造方法,否则无法通过编译。因此,建议在定义父类时,如写了构造方法,最好写上一个无参数的构造方法。

第 18 题:本题考查类的继承。选项 A 中的 m 仅在父类的方法参数中定义,无法给变量 i 赋值;选项 B 中的 b 是非静态成员变量,在静态的 main 方法中无法直接给 i 赋值;父类定义的 a 是私有的,无法被子类继承,所以选项 C 也不对。选项 D 中的方法 change 是父类定义的公共方法,可以被子类继承和调用。因此,本题选 D。

第 23 题:本题考查类的构造方法。程序中第 2 行为构造方法,第 5 行为方法名,Test 有 void 修饰,它不是构造方法,可作为实例方法。程序编译能通过。运行时,第 9 行实例化 Test 对象,调用第 2 行的构造方法,输出 3;然后执行第 9 行调用第 5 行的方法,输出 2;最后执行 11 行,输出 1。因此,本题选 C。

第 24 题:本题考查子类实例化时的过程,具体见例 4-1。

第 27 题:本题考查泛型类的实例化。本题中的类 C4 泛型参数是 C1&C2,其中 C1 是类,C2 是接口。因此,对 C4 实例化时,它的泛型参数必须具有继承类 C1 且实现接口 C2 的特性,而类 C3 具有这种特性。所以,本题选 C。

17.3 程序阅读题

1. int
2. (1) abstract
   (2) getName
   (3) extends
   (4) String major
   (5) getMajor
3. 32
4. resultOne=0
   resultTwo=48
   resultThree=20
5. Pine
   Tree
   Oops
6. My Func
7. 输出结果为:2

8. hi!
   I am Tom
   How do you do?
9. 第 2 行改为 final double PI=3.14；
   第 5 行的 extends 改为 implements
   第 10 行的方法声明增加一个 public，改为 public double area()
10. strFoo.getX=Hello!
    douFoo.getX=1.0
11. 15
    11
    10
12. 2
    2
    1
13. i of methodI():21
    i of Test11:3
    i of Test1:2
14. 第 26 行输出 11
    第 27 行输出 5
    第 28 行编译出错，因为父类 ParentClass 中没有定义 multi(int,int)
    第 29 行输出 0
    第 30 行编译出错，因为父类中没有定义 y
    第 31 行输出 11
    第 32 行输出 5
    第 33 行输出 25
    第 34 行输出 1
    第 35 行输出 2

17.4 编 程 题

1. 参考代码如下：

```
public class Circle {
    double radius;
    public Circle(){
        radius=0;
    }
    public Circle(double r){
        radius=r;
    }
    public double getRadius(){
```

```java
        return radius;
    }
    public double getPerimeter(){
        return 2 * Math.PI * radius;
    }
    public double getArea(){
        return Math.PI * radius * radius;
    }
    public void disp(){
        System.out.println("圆的半径为"+getRadius()+" 周长为"+getPerimeter()+" 面积为"+getArea());
    }
}

public class Cylinder extends Circle{
    double height;
    public Cylinder(double r,double h){
        radius=r;
        height=h;
    }
    public double getHeight(){
        return height;
    }
    public double getCylinderArea(){
        return 2 * Math.PI * radius * (radius+height);
    }
    public double getVol(){
        return Math.PI * radius * radius * height;
    }
    public void disVol(){
        System.out.println("圆柱体的体积为"+getVol());
    }
}
```

2. 参考代码如下：

```java
interface Shape{
    public abstract double getArea();
}
class Circle implements Shape{
    private int r;
    Circle(int r){
        this.r=r;
    }
    public  double getArea(){
        return 3.14 * r * r;
```

```java
        }
    }
class TestCircle{
    public static void main(String args[]){
        Circle c=new Circle(5);
        System.out.println("圆的面积为:"+c.getArea());
    }
}
```

3. 参考代码如下：

```java
interface shape{
    double area();
}
class Triangle implements shape{
    private double a, b, c;
    public Triangle(double a, double b, double c){
        this.a=a;
        this.b=b; this.c=c;
    }
    public double area(){
        double p= (a +b +c) / 2;
        return Math.sqrt(p * (p -a) * (p-b) * (p-c));
    }
}
class Circle implements shape{
    private double r;
    public Circle(double r){
        this.r=r;
    }
    public double area(){
        return Math.PI * r * r;
    }
}
class Rectangle implements shape{
    private double width, height;
    public Rectangle(double j, double k){
        width=j; height=k;
    }
    public double area(){
        return width * height;
    }
}
public class TestShape {
    public static void main(String args[]){
        shape s[]=new shape[3];
```

```java
            s[0]=new Triangle(3,4,5);
            s[1]=new Circle(3.5);
            s[2]=new Rectangle(3.5,4.0);
            for(int i=0;i<s.length;i++)
                System.out.println(s[i].area());
    }
}
```

4. 参考代码如下：

```java
interface DataStructure{
    boolean isEmpty();
    boolean isFull();
    void addElement(Object obj);
    Object removeElement();
}
class MyQueue implements DataStructure{
    private int idx=0,size;
    private Object[] data;
    public MyQueue(int size){
        data=new Object[size];
        this.size=size;
    }
    public boolean isEmpty(){
        if(idx==0)
            return true;
        else
            return false;
    }
    public boolean isFull(){
        if(idx==size)
            return true;
        else
            return false;
    }
    public void addElement(Object obj){
        data[idx++]=obj;
    }
    public Object removeElement(){
        Object obj=data[0];
        for(int i=0;i<idx-1;i++){
            data[i]=data[i+1];
        }
        idx--;
        return obj;
    }
```

```
}
public class TestDataStructure{
    public static void main (String[] args)    {
        DataStructure ds=new MyQueue(10);
        for(int i=0;i<10;i++)
            ds.addElement(""+i);
        while(!ds.isEmpty())
            System.out.println ((String)ds.removeElement());
    }
}
```

5. （1）参考代码如下：

```
abstract public class Animal {
    public String name;
    public int age;
    public double weight;
    public void showInfo( ){
        System.out.println("动物名为"+name+",年龄为"+age+"岁,重量为"+weight);
    }
    abstract public void move();
    abstract public void eat();
}
```

（2）参考代码如下：

```
public class Bird extends Animal{
    public Bird(String name,int age,double weight){
        this.name=name;
        this.age=age;
        this.weight=weight;
    }
    public void showInfo( ){
        System.out.println("鸟名为"+name+",年龄为"+age+"岁,重量为"+weight);
    }
    public void move(){
        System.out.println(name+"用翅膀在天空上飞!");
    }
    public void eat(){
        System.out.println(name+"喜欢吃虫子!");
    }
}
```

（3）参考代码如下：

```
public class TestAnimal {
    public static void main(String args[]){
        Animal bird=new Bird("麻雀",1,0.3);
```

```java
        bird.showInfo();
        bird.move();
        bird.eat();
    }
}
```

6. 参考代码如下：

```java
public class People {
    public String name;
    public String sex;
    public String bothnum;
    public People(){};
    public String printInfo(){
        return"姓名:"+' '+name+'\n'+"性别:"+' '+sex+'\n'+"出生年月:"+' '+bothnum;
    }
}
class Teacher extends People{
    public String school;
    public int workID;
    public String printInfo(){
        return super.printInfo()+'\n'+"学校:"+' '+school+'\n'+"工号:"+' '+workID;
    }
}
class Student extends People{
    public String school;
    public int Id;
    public String discipline;
    public String grade;
    public String classes;
    public String printInfo(){
        return super.printInfo()+'\n'+"学校:"+' '+school+'\n'+"学号:"+' '+Id+'\n'+"专业:"+' '+discipline+'\n'+"年级:"+' '+grade+'\n'+"班级:"+' '+classes;
    }
}
class TestPerson{
public static void main(String args[]){
    Teacher t=new Teacher();
    t.name="Tom";
    t.sex="男";
    t.bothnum="19820808";
    t.school="浙江工业大学";
    t.workID=2008010408;
    System.out.println(t.printInfo());
    Student s=new Student();
    s.name="Jack";
```

```
        s.sex="男";
        s.bothnum="19920305";
        s.Id=2011266302;
        s.school="浙江工业大学";
        s.discipline="软件工程";
        s.grade="大一";
        s.classes="2班";
        System.out.println(s.printInfo());
    }
}
```

7. 略。

8. (1) 参考代码如下：

```
interface AreaInterface {
    public static final double pai=Math.PI;
    public abstract double area();
}
```

(2) 参考代码如下：

```
public class Rectangle implements AreaInterface{
    private double x;
    private double y;
    public Rectangle(double x,double y){
        this.x=x;
        this.y=y;
        System.out.println("长方形的长为："+x+"宽为："+y);
    }
    public double area(){
        return x * y;
    }
    public String toString(){
        return ("长方形的面积为："+this.area());
    }
}
```

(3) 参考代码如下：

```
class TestArea{
    public static void main(String args[]){
        Rectangle r1=new Rectangle(10.0,20.0);
        System.out.println(r1.toString());
    }
}
```

9. 略。

10. 略。

11. 参考代码如下：

```java
class Outer {
    String name;
    int i;
    public Outer(){
        name="Outer";
        i=20;
    }
    class Inner{
        String name;
        int i;
        public Inner(){
            name="Inner";
            i=10;
        }
        public void printInfo(){
            System.out.println("外部类:"+'\n'+"name: "+Outer.this.name+"   "+"i: "+Outer.this.i);
            System.out.println("内部类: "+'\n'+"name:"+name+" "+"i: "+i);
        }
    }
}
class TestOuter{
    public static void main(String args[]){
        Outer o=new Outer();
        Outer.Inner i=o.new Inner();
        i.printInfo();
    }
}
```

第 18 章　数组、字符串和枚举参考答案

18.1　判　断　题

1. 错　2. 对　3. 对　4. 对　5. 对　6. 错　7. 错　8. 错

【解析】

第 6 和 8 题：这两题考查枚举类型的定义。在 Java 中，创建枚举类型的主要目的是为了定义一些枚举常量。枚举类型不能定义方法，也不能通过 new 创建实例对象。

18.2　选　择　题

1. A　2. D　3. A　4. C　5. A　6. C　7. D　8. C　9. B
10. C　11. B　12. C　13. A　14. A　15. A　16. D　17. BD　18. B
19. A　20. B　21. D　22. B　23. C　24. B　25. A　26. B　27. B

【解析】

第 5 题：本题考查数组的创建和长度。数组一旦被创建，其长度无法被改变。数组的 length 属性是 final 修饰的，不能改变它的值。因此，本题选 C。

第 6 题：本题考查字符串 String 方法的应用。字符串 String 是一种内容不可变对象的类，一旦创建后，不能改变它的值。本题第 4 行表达式语句执行的结果是 DE，但字符串变量 str 本身的值没变，还是 ABCDE；同理，第 5 行表达式语句执行的结果是 ABCDEXYZ，但字符串变量 str 本身的值还是 ABCDE。本题选 C。

第 7 题：本题考查二维数组的创建。二维数组可以看成是一维数组的数组，可以直接为每一维分配空间，也可以从最高维开始，分别为每一维分配空间，创建以数组为元素的数组，即二维数组的每一行可以具有不同的列数。因此，选项 B 是合法的，而选项 D 由于最高维没定义，不合法。

第 9 题：本题考查数组和字符串作为参数传递给方法调用。数组也是引用类型，作为参数传递给方法调用时，在方法中可修改数组中的元素值。而字符串 String 是特殊的引用类型，不能改变它本身的值。因此，本题选 B。

第 14 题：本题考查字符串 String 对象的内存存储机制和"=="运算。第 3～6 行的 s1、s2 和 s3 的引用地址都是字符串池中"123456"的内存地址；第 7 行，由于"+"运算中有一个操作数是变量 a0，无法在编译中直接完成"+"运算，所以 s4 的引用地址不在字符串池中；第 8 行，用 new 新创建了一个字符串对象，s5 的引用地址也不在字符串池中；第 9 行，s5 调用了调用 intern 方法，该方法的作用是如果池已经包含一个等于此 String 对象的字符串，则返回池中的字符串，否则，将此 String 对象添加到池中，并返回此 String 对象的引用，所以 s6 的引用地址也是字符串池中"123456"的内存地址。对引用对象做"=="运算时，若引用对象的地址相同，则返回 true，否则返回 false。因此，本题选 A。

第 15 题：本题考查 StringBuffer 类的方法 length 和 capacity。StringBuffer 类的方法 length 是返回 StringBuffer 对象中字符串的长度，而方法 capacity 是返回 StringBuffer 对象可存放字符串的容量大小。声明语句"StringBuffer buf1＝new StringBuffer(20);"的内容为空字符串，容量大小为 20。因此，本题选 A。

第 18 题：本题考查对象数组的创建。语句"String[] s＝new String[10];"执行后，创建了一个字符串数组对象 s，数组长度为 10，但数组中的所有内容都是 null。选项 A 的 s[10]超出数组的范围了，选项 D 数组的长度应该是 s.length。所以本题选 B。

第 20 题：本题考查"＝＝"运算和字符串 equals 方法。选项 A 中"＝＝"运算的左右两边的操作数都是字符串常量，返回 true；选项 B，equals 方法比较的是不同的对象，返回 false；选项 C 和 D，equals 方法比较的都是字符串对象，字符串对象中的内容相同，都返回 true。因此，本题选 B。

18.3　程序阅读题

1. (1) arraySum 方法的功能是返回整型数组中所有元素的和。
   (2) 运行结果是 100。
2. 春季，夏季，秋季，冬季，
3. java
   iavaC
   Yava
4. (1) 结果为(8.0,12.0)
   (2) 数据格式错！

18.4　编　程　题

1. 略。
2. 参考代码如下：

```
import java.util.*;
class Set extends HashSet{
  public Set intersection(Set set){          //求交集
    Set newSet=new Set();
    for(Object ob:this){
      if(set.contains(ob)){
        newSet.add(ob);
      }
    }
    return newSet;
  }
  public Set union(Set set){                 //求并集
    this.addAll(set);
```

```java
            return this;
        }
    }
    public class SetTest {
        public static void main(String args[])    {
            Set s1=new Set();
            Set s2=new Set();
            for(int i=0;i<5;i++)    {
                s1.add((int)(Math.random() * 10));
            }
            for(int i=0;i<8;i++)    {
                s2.add((int)(Math.random() * 10));
            }
            System.out.println(s1);
            System.out.println(s2);
            System.out.println("交集:"+s1.intersection(s2));
            System.out.println("交集:"+s1.union(s2));
        }
    }
```

3. 参考代码如下:

```java
class Matrix{
    int m[][]=new int[5][5];
    public Matrix()    {
        for(int i=0;i<5;i++){
            for(int j=0;j<5;j++){
                m[i][j]=(int)(Math.random() * 10);
            }
        }
    }
}
public class TestMatrix {
    public static void main(String args[])    {
        Matrix a=new Matrix();
        Matrix b=new Matrix();
        Matrix c=new Matrix();
        for(int k=0;k<5;k++)    {
            for(int i=0;i<5;i++){
                for(int j=0;j<5;j++)    {
                    c.m[k][i]+=(a.m[k][j]) * (b.m[j][i]);
                }
                System.out.print(c.m[k][i]+" ");
            }
            System.out.println();
```

    }
}

4. 略。

5. 参考代码如下：

```java
public class XyArray {
    public static void main(String args[]){
        int n=5;
        int array[][]=new int[n][n];
        for(int i=0;i<n;i++)    {
            for(int j=0;j<n;j++){
                if(i==j)
                    array[i][j]=1;
                else
                    array[i][j]=0;
                System.out.print(array[i][j]);
            }
            System.out.println();
        }
    }
}
```

6. 参考代码如下：

```java
import java.util.*;
public class SortArray {
    public static void main(String args[])     {
        int a[]={20,10,50,40,30,70,60,80,90,100};
        Arrays.sort(a);
        for(int i=a.length-1;i>=0;i--)
            System.out.print(a[i]+" ");
    }
}
```

7. 参考代码如下：

```java
public class TestArraySum {
    public static void main(String args[])     {
        TestArraySum t=new TestArraySum();
        int a[]=new int[10];
        for(int i=0;i<10;i++){
            a[i]=(int)(Math.random() * 9+11);
        }
        System.out.print(t.arraySum(a));
    }
    public int arraySum(int a[]){
        int sum=0;
```

```java
            for(int b:a)
                sum+=b;
            return sum;
        }
    }
```

8. 参考代码如下：

```java
import java.util.Arrays;
public class WordSort {
    public static void main(String args[])    {
        String s[]={"hello","world","welcome","hi","hey"};
        String minst=null;
        for(int i=0;i<s.length;i++)    {
            for(int j=i+1;j<s.length;j++){
                if(s[i].compareTo(s[j])>0){ //比较两个字符串的大小
                    minst=s[i];
                    s[i]=s[j];
                    s[j]=minst;
                }
            }
            System.out.print(s[i]+" ");
        }
    }
}
```

9. 略。

10. 参考代码如下：

```java
public class StatisticsWord {
    public static void main(String args[]){
        String s[]={"we","will","word","what","and","two","out",
            "I","hope","you","can","me","please","accept","my","best"};
        int a,b,c;
        a=b=c=0;
        for(int i=0;i<s.length;i++){
            /*统计以字母 w 开头的单词数*/
            if((s[i].substring(0,1)).equals("w")){
                ++a;
            }
            /*统计单词中含"or"字符串的单词数*/
            if(s[i].indexOf("or")>=0){
                ++b;
            }
            /*统计长度为 3 的单词数*/
            if(s[i].length()==3){
```

```
                ++c;
            }
        }
        System.out.print("字母 w 开头的单词数为："+a+"\n"+"单词中含有"or"的单词数为：
        "+b+"\n"+"单词长度为 3 的单词数为："+c);
    }
}
```
11. 略。

第 19 章 Java 常用类及接口参考答案

19.1 判 断 题

1. 对 2. 对 3. 对 4. 对 5. 对

19.2 选 择 题

1. B 2. C 3. B 4. A 5. AB

【解析】

第 1 题：由于 Math.random()产生的随机数范围为[0.0,1.0)，不能取得 1.0，所以选项 A 产生的整数为[20,998]，选项 B 产生的整数为[20,999]；选项 C 和 D 中的(int)Math.random() * 999 和(int)Math.random() * 980 的值为 0。本题选 B。

第 4 题：Set 不允许添加重复的元素，但允许对 set 对象中的元素值修改后相等。如果加入一个重复元素，add 方法将返回 false。本题选 B。

19.3 程序阅读题

1. 3
   "3"is an element of vs
   2
2. 120

19.4 编 程 题

1. 参考代码如下：

```
import java.text.SimpleDateFormat;
import java.util.Date;
public class ShowDateTime {
    public static String dateTimeToString(Date time){
        SimpleDateFormat formatter;
        formatter=new SimpleDateFormat ("yyyy-MM-dd HH:mm:ss");
        return formatter.format(time);
    }
    public static String dateToString(Date time){
        SimpleDateFormat formatter;
        formatter=new SimpleDateFormat ("yyyy年MM月dd日");
```

```
            return formatter.format(time);
        }
        public static void main(String args[]){
            System.out.println("今天日期是"+dateToString(new Date()));
            System.out.println("当前时间是"+dateTimeToString(new Date()));
        }
    }
```
2. 略。
3. 略。

第 20 章 异常处理参考答案

20.1 选 择 题

1. D 2. C 3. D 4. B 5. C 6. D 7. D 8. B 9. C 10. C

【解析】

第 8 题：本题考查 try-catch-finally 语句的执行过程。若 try 块中的某一个语句产生异常，则 try 中剩下的语句就不执行了，产生的异常对象由 catch 捕获处理，若没被 catch 捕获，则将产生的异常对象上传至调用该方法的语句。对 finally 语句来说，除非在 try 或 catch 语句中有强制退出程序的语句 System.exit(0)出现，否则不管有没有产生异常，也不管异常有没有被捕获处理，都要执行。因此，本题中若 unsafe()有异常，则输出 234，若没有异常，则输出 134。

第 9 和 10 题：这两题考查数组越界的异常。对数组的访问出现越界的情况，编译都是没问题，但在运行时会出现数组下标越界的异常。

20.2 程序阅读题

1. 程序中标记序号的执行顺序是：1,4,5,6,2,3
2. （1）cmn
   （2）bcdn
3. 捕获 ArithmeticException 异常
   没有 13 月份！
4. 边长分别为 5,5,10，不能构成三角形
5. 略
6. 张三的工资为 1000 元。
   异常：工资不能为负！
7. 求 12 的因子个数
   求 12 的因子个数有误
   结束

20.3 编 程 题

1. 参考代码如下：

```
import java.util.Arrays;
public class StuScoreSort1 {
    public static void main(String[] args) {
```

```
        int[] a=new int[10];
        try {
            int k=0;
            for(int i=0; i<10; i++) {
                try{
                    a[i]=Integer.parseInt(args[i]);
                    k++;
                }catch (Exception e){
                    System.out.println("请输入正确的成绩!");
                }
            }
            if(k<10){
                throw new Exception();
            }
            Arrays.sort(A.;                        //按从小到大排序
            for(int i=a.length-1;i>=0;i--){        //按高到低输出成绩
                System.out.print(a[i] +" ");
            }
        }
        catch(Exception e){
            System.out.println("请输入至少 10 个成绩");
        }
        finally{
            System.out.println("结束");
        }
    }
}
```

2. 参考代码如下：

```
class bankException extends Exception{           //自定义异常类
    public bankException(String s){
        super(s);
    }
}
public class Bank {
    double balance;
    public Bank(double balance){
        this.balance=balance;
    }
    void deposite(double storeMoney) {            //存款
        balance+=storeMoney;
    }
    void getbalawal(){                            //查询余额
        System.out.print(balance);
    }
```

```java
        void withdrawa(double getMoney) {            //取款
            try{
                if(balance<getMoney)    {
                    throw new bankException("余额:"+balance+"不能取款!");
                }
                else{
                    balance-=getMoney;
                    System.out.println("取款成功!");
                }
            }catch(bankException e){
                System.out.println(e.toString());
            }
        }
        public static void main(String args[])    {
            Bank b=new Bank(1000);
            b.deposite(500);                    //存 500
            b.withdrawa(1000);                  //取 1000
            b.withdrawa(1000);                  //再取 1000
        }
}
```

3. 参考代码如下:

```java
class EmptyStackException extends Exception{
    public EmptyStackException(String s)    {
        super(s);
    }
}
class FullStackException extends Exception{
    public FullStackException(String s){
        super(s);
    }
}
```

4. 参考代码如下:

```java
class MyStack<T>{
    int size;
    T element[];
    int i=-1;
    public MyStack(int size){
        this.size=size;
    }
    public boolean isEmpty(){
        return i==-1;
    }
    public boolean isFull(){
```

```java
            return i==size-1;
        }
        public void push(T s)    {
            try{
                if(isFull())
                    throw new FullStackException("堆栈满!");
                else{
                    i++;
                    element[i]=s;
                }
            }catch(FullStackException fe)    {
                System.out.print(fe.toString());
            }
        }
        public void pop()    {
            try{
                if(isEmpty())
                    throw new EmptyStackException("堆栈空!");
                else{
                    element[i]=null;
                    i--;
                }
            }catch(EmptyStackException e)    {
                System.out.print(e.toString());
            }
        }
    }
    public class TestMyStack {
        public static void main(String args[]){
            MyStack<String>b=new MyStack<String>(5);
            b.element=new String[5];
            b.push("a");
            b.push("b");
            b.push("c");
            for(int j=b.i;j>=0;j--)    {
                System.out.print(b.element[j]+" ");
            }
            System.out.println();
            b.pop();
            for(int j=b.i;j>=0;j--)
                System.out.print(b.element[j]+" ");
        }
    }
```

第 21 章 流和文件参考答案

21.1 判 断 题

1. 对 2. 对 3. 对 4. 对 5. 对

21.2 选 择 题

1. D 2. B 3. C 4. A 5. D 6. D 7. A 8. B
9. C 10. B 11. B 12. C

【解析】

第 4 题：本题考查 FileInputStream 和 RandomAccessFile 读取文件内容的方法。选项 D 改为以下代码，则可满足要求。

```
RandomAccessFile in=new RandomAccessFile("file.txt","r");
in.seek(9);
int c=in.readByte();
```

第 12 题：本题考查 FileOutputStream 的构造方法。本题创建 FileOutputStream 的实例对象，其中第一个参数 data.txt 为文件名，第二个参数为 true。如果文件 data.txt 不存在，但有写文件的权限，也不一定会抛出 IOException 异常，而会新建文件 data.txt；如果文件 data.txt 存在，则从文件的末尾开始添加新内容。若创建 FileOutputStream 的实例对象时，第二个参数为 false，或者无第二个参数，当文件存在时，将覆盖掉文件中原有的内容。因此，本题选 C。

21.3 程序阅读题

1. 第 9 行表示为 path 创建了一个 File 对象，对象引用名是 f
   第 12 行表示对象 f 是否为文件
   程序功能是打印 C 盘下所有文件的绝对路径名
2. 第 4 行语句为：Test t＝new Test();
   第 9 行含义：为 test.txt 创建了 FileInputStream 对象，对象引用名是 din
   第 11 行含义：捕获 IOException 异常，对象引用名为 ioe
   运行结果为
   two
   0
3. (1) java.io.*

(2) new FileReader("Test.txt")

(3) br.readLine()

(4) new FileWriter("myTest.txt")

(5) catch(IOException e)

4. 文件没找到

退出

−1

5. (1) new FileReader(articleName)

(2) br.readLine()

(3) sentence.indexOf(word,index+word.length())

(4) WordCount w=new WordCount()

21.4 编 程 题

1. 参考代码如下：

```
import java.io.*;
public class ReadText {
    public static void main(String args[]){
        try{
            BufferedReader br=new BufferedReader(new FileReader("myText.txt"));
            String s=null;
            while((s=br.readLine())!=null){        //按行读取文件内容
                System.out.println(s);
            }
            br.close();
        }catch(Exception e){
            System.out.print(e.toString());
        }
    }
}
```

2. 参考代码如下：

```
import java.io.File;
public class CheckFile {
    public static void main(String args[]){
        File f=new File("myText.txt");
        if(f.exists())                             //判断文件是否存在
            System.out.print(f.length());
    }
}
```

3. 参考代码如下：

```java
import java.io.*;
public class WriteLog{
    public static void main(String []args){
        String f="d:/log.txt";
        String str="";
        try{
            BufferedReader keyIn=new BufferedReader(new InputStreamReader(System.in));
            PrintWriter bw=new PrintWriter(new FileWriter(f));
            System.out.println("Please input file text:");
            while(!(str=keyIn.readLine()).equals("quit#")){
                bw.println(str);
            }
            bw.close();
        }catch(IOException e){ }
    }
}
```

4. 参考代码如下：

```java
import java.io.*;
public class WordStatistic {
    public static void main(String[] args) throws IOException{
        FileInputStream is=new FileInputStream("word.txt");
        int a;
        int capital=0;
        int lowercase=0;
        while((a=is.read())!=-1){
            if(a>=65&&a<=91){
                capital++;
            }
            if(a>=97&&a<=125){
                lowercase++;
            }
        }
        System.out.println("大写字母有"+capital+"个");
        System.out.println("小写字母有"+lowercase+"个");
    }
}
```

5. 参考代码如下：

```java
import java.util.*;
import java.io.*;
public class Digital {
    public static void main(String args[]){
```

```
            Random r=new Random();
            int n=10;
            int num[]=new int[n];
            for(int i=0;i<n;i++){
                num[i]=r.nextInt(40)+10;
            }
            Arrays.sort(num);
            try{
              BufferedWriter bw=new BufferedWriter(new FileWriter("n.txt"));
              for(int j=n-1;j>=0;j--) {
                  bw.write(num[j]);
                  System.out.print(num[j]+" ");
              }
              bw.close();
            }
            catch(Exception e){
                System.out.println(e.toString());
            }
        }
    }
```

6~14 略。

第22章 图形用户界面编程参考答案

22.1 判断题

1. 错 2. 错 3. 对 4. 错 5. 对 6. 对 7. 错 8. 对 9. 对
10. 对 11. 错 12. 错 13. 对

【解析】

第1题：本题考查容器组件。JButton 不是容器组件。

第2题：本题考查布局管理器。BorderLayout 是 JFrame 的默认布局管理器，而 FlowLayout 是面板 JPanel 的默认布局管理器。

第4题：本题考查面板 JPanle 的特性。一个 JPanel 可以加入到另一个 JPanel 中，形成嵌套布局。

第7题：本题考查适配器类。Java 中的监听器接口中若有两个或两个以上方法的，该监听器接口才有对应的适配器类。

第11题：本题考查事件监听处理的机制。如果一个组件注册多个监听器，事件会被所有的监听器同时处理。

第12题：本题考查 JList 的事件处理。单击 JList 组件会产生 ItemEvent 事件，并由监听器 ItemListener 处理。

22.2 选择题

1. B 2. C 3. D 4. BD 5. C 6. B 7. B 8. C 9. B
10. A 11. B 12. D 13. C 14. B 15. C 16. D

【解析】

第12题：本题考查 BorderLayout 布局的特点。采用 BorderLayout 布局时，添加到 NORTH 区的两个按钮仅显示后面添加的那个按钮。

第13题：本题考查鼠标事件的处理器接口。处理鼠标事件的事件监听器接口包括 MouseListener、MouseMotionListener 和 MouseWheelListener。

22.3 程序阅读题

1. (1) extends JFrame
   (2) 调用父类 JFrame 的构造方法，使得界面的标题为 Concentric
   (3) 设置界面窗口的大小为 200×100
   (4) 使得界面显示出来

2. 本题的运行界面如图 22-1 所示。
3. 第 11 和 13 行的 container.put 改为 container.add
   第 15 和 16 行的 addItemListener 改为 addActionListener
   第 25 行的 itemStateChanged 改为 ActionPerformed
4. (1) javax.swing.*
   (2) getContentPane()
   (3) c.setLayout(new BorderLayout())
   (4) setVisible(true)

图 22-1 运行界面(一)

本题添加完整的代码后,编译、运行的界面如图 22-2 所示。
5. 本题没有编译错误,执行结果的界面如图 22-3 所示。

图 22-2 运行界面(二)　　　图 22-3 运行界面(三)

6. (1) java.awt.event.*
   (2) implements ActionListener
   (3) JFrame f=new JFrame("按钮测试");
   (4) actionPerformed(ActionEvent e)
   (5) e.getSource()==ok
7. (1) implement ActionListener
   (2) number.addActionListener(this)
   (3) Integer.parseInt(s)
   (4) info="输入错误!"
   (5) wordInfo.setText(info)
8. (1) import java.io.*
   (2) extends JFrame implements ActionListener
   (3) c.add(text,"Center")
   (4) actionPerformed(ActionEvent e)
   (5) new FileWriter("mytext.txt")

22.4　编　程　题

略。

第 23 章　多线程参考答案

23.1　判　断　题

1. 对　2. 对　3. 错　4. 对　5. 错　6. 错　7. 对　8. 错　9. 错
10. 对　11. 对　12. 错　13. 对　14. 对　15. 错

【解析】

第 3 题：本题考查启动线程的方法。线程采用 start() 方法启动，线程启动后进入就绪状态等待系统的调度，可体现线程并发执行的特性。若线程对象直接调用 run() 方法，则直接执行 run() 方法中的语句，这与普通的方法调用无异。

第 5 题：本题考查 sleep 方法。调用 sleep 方法结束后，线程重新进入就绪状态等待系统的调用。

第 6 题：本题考查多线程执行的并发性。多个线程的运行顺序不是按线程启动的顺序进行的，而是并发执行的。

第 8 题：本题考查唤醒线程的方法。notifyAll() 方法会激活在等待集中的所有线程。

第 9 题：本题考查线程优先级和设置方法。每个 Java 线程的优先级都设置在常数 1 到常数 10 之间，默认的优先级设置为常数 5。

第 15 题：本题考查 sleep() 方法的特点。线程使用 sleep() 方法休眠结束后，不需要 notify() 方法唤醒，而是会自动进入就绪状态等待系统的调用。

23.2　选　择　题

1. B　2. D　3. C　4. D　5. B　6. B　7. C　8. B
9. A　10. D　11. B　12. D　13. C　14. B

【解析】

第 12 题：本题考查线程并发执行的特性。线程 t1 和 t2 还未启动执行前，main 方法的主线程已经在打印了，此时可能输出 Yes Yes，也有可能是 t1 已启动，t2 未启动，main 方法的主线程打印 No Yes，也有可能是打印 No No。若在多核 CPU 的机器上执行比较明显地会出现各种情况。总之，输出结果是无法确定的。

23.3　程序阅读题

1. 在屏幕上每隔 1s 输出一个数字，最后输出 1 2 3 4 5
2. (1) implements Runnable
　　(2) (char)('a'+i)
　　(3) Thread.sleep(1000)

  (4) new Thread(new Test())

  (5) t. start()

3. (1) new Thread(new T1())

  (2) t1. setName("线程 1")

  (3) t1. start()

  (4) implements Runnable

  (5) Thread. sleep((int)(Math. random() * 5000))

4. (1) extends Thead

  (2) String str,int speed

  (3) a1=new Animal("rabit",100)

  (4) a2=new Animal("turtle",20)

  (5) a2. setPriority(Thread. MAX_PRIORITY)

5. (1) synchronized

  (2) available==false

  (3) wait()

  (4) notifyAll()

  (5) available==true

6. 本题有两种修改方法：

(1) 如果采用 implements Runnable 实现线程,则修改的地方有以下 3 处：

① 第 3～5 行写入构造方法,并且将第 3 行的 str. length()改为 str. length,修改后的代码如下：

```
public Test() {
    for (int i=0; i<str.length; i++) {
        str[i]=i+"-";
    }
}
```

② 第 8 行的 str. length()改为 str. length。

③ 第 9 行的"sleep(1000);"改为"Thread. sleep(1000);"。

注意：从程序正常运行的角度,第 18 行 t. run();不用修改。

(2) 如果采用 extends 实现该线程程序,则修改的地方有以下 3 处。

① 第 1 行的 implements Runnable 修改为 extends Thread。

② 第 3～5 行写入构造方法,并且将第 3 行的 str. length()改为 str. length,修改后的代码如下：

```
public Test() {
    for (int i=0; i<str.length; i++) {
        str[i]=i+"-";
    }
}
```

③ 第 8 行的 str. length()改为 str. length。

7. 第 4 行修改为 new Thread(b).start();
   第 12 行修改为 Thread.sleep(1000);
   程序修改后,经编译和运行,输出结果为
   第 1 次
   第 2 次
8. 1010
   4010

23.4 编 程 题

略。

第 24 章 网络编程参考答案

24.1 判 断 题

1. 对 2. 对 3. 错 4. 对 5. 对

【解析】
第 3 题：本题考查关闭 Socket 的方法。服务器执行 Socket 的 close()方法来关闭连接。

24.2 选 择 题

1. B 2. C 3. D 4. B 5. C 6. B 7. D 8. D 9. A 10. B

24.3 程序阅读题

1. (1) server＝new ServerSocket(6000,50)
   (2) output＝new ObjectOutputStream(socket.getOutputStream());
   (3) input＝new ObjectInputStream(socket.getInputStream());
   (4) output.close();
   (5) input.close();
2. (1) Runnable
   (2) ServerExample
   (3) run()
   (4) ServerSocket
   (5) ServerSocket
   (6) Socket
   (7) Thread
   (8) t
3. (1) new DatagramSocket(8000)
   (2) dSocket.receive(inPacket)
   (3) new String(inPacket.getData())
   (4) new DatagramPacket(outBuffer,outBuffer.length,cAddr,cPort)
   (5) dSocket.send(outPacket)

24.4 编 程 题

略。

第 25 章 数据库编程参考答案

25.1 选 择 题

1. D 2. C 3. ABCD 4. A 5. AB 6. A
7. AC 8. A 9. ABCD 10. ACD 11. ABCD 12. ACD
13. A 14. D 15. C

25.2 编 程 题

略。

第26章 XML及程序打包参考答案

26.1 判 断 题

1. 对 2. 对 3. 对 4. 对 5. 对

26.2 选 择 题

1. B 2. ABCD 3. A 4. CD 5. C 6. B 7. ABC 8. A

参 考 文 献

[1] 雍俊海. Java 程序设计. 北京:清华大学出版社,2007.
[2] 雍俊海. Java 程序设计习题集. 北京:清华大学出版社,2007.
[3] 郎波. Java 语言程序设计. 北京:清华大学出版社,2005.
[4] 施霞萍等. Java 程序设计教程. 北京:机械工业出版社,2006.
[5] Java 语言程序设计习题集. http://wenku.baidu.com/view/2f0a0e1fc5da50e2524d7f58.html.